MULTIVARIATE FREQUENCY ANALYSIS OF HYDRO-METEOROLOGICAL VARIABLES

MULTIVARIATE FREQUENCY ANALYSIS OF HYDRO-METEOROLOGICAL VARIABLES

A COPULA-BASED APPROACH

FATEH CHEBANA

Water, Earth and Environment Centre, National Institute of Scientific Research, University of Quebec, Quebec, Canada

ELSEVIER

Elsevier
Radarweg 29, PO Box 211, 1000 AE Amsterdam, Netherlands
The Boulevard, Langford Lane, Kidlington, Oxford OX5 1GB, United Kingdom
50 Hampshire Street, 5th Floor, Cambridge, MA 02139, United States

Notices
Knowledge and best practice in this field are constantly changing. As new research and experience broaden our understanding, changes in research methods, professional practices, or medical treatment may become necessary.

Practitioners and researchers must always rely on their own experience and knowledge in evaluating and using any information, methods, compounds, or experiments described herein. In using such information or methods they should be mindful of their own safety and the safety of others, including parties for whom they have a professional responsibility.

To the fullest extent of the law, neither the Publisher nor the authors, contributors, or editors, assume any liability for any injury and/or damage to persons or property as a matter of products liability, negligence or otherwise, or from any use or operation of any methods, products, instructions, or ideas contained in the material herein.

ISBN: 978-0-323-95908-7

For Information on all Elsevier publications
visit our website at https://www.elsevier.com/books-and-journals

Publisher: Candice G. Janco
Acquisitions Editor: Maria Elekidou
Editorial Project Manager: Andrae Akeh
Production Project Manager: Rashmi Manoharan
Cover Designer: Harris Greg

Typeset by MPS Limited, Chennai, India

Contents

Acknowledgments

I would like to acknowledge all those who have contributed to the preparation and completion of this book. I am grateful to my colleagues and friends, from INRS and other institutions from Canada and abroad, who gave comments directly or indirectly related to the topic of this book, in particular, the STAHY (International Commission on Statistical Hydrology) members. I have been greatly influenced by the discussions and support of Abdelhameed El-Shaarawi, founder and former Editor-in-Chief of Environmetrics.

I would like to express my gratitude to my students and team members, especially Dorsaf Goutali, for all of her help during the last year.

All efforts, commitments, and patience from Elsevier team are greatly appreciated.

I'm grateful to my family, especially my wife, Narimane, and our kids, for their patience, support, and encouragement over the years.

I express my sincere thanks to each person concerned.

I dedicate this book to my parents, Achour and Laatra!!

Fateh Chebana
Quebec City, Canada

CHAPTER

1

Introduction

1.1 Context

Extreme hydro-meteorological events, including floods, droughts, and storms, may have serious economic and social consequences. Given the importance of dealing with such adverse events, several studies have focused specifically on each of these events. In water resource management, one of the challenging topics is related to droughts, whereas the design of hydraulic structures is based on floods. An accurate estimation of the risk caused by these events is essential. In this regard, hydrological frequency analysis (HFA), as a set of statistical methods and techniques, is commonly considered. HFA is mainly composed of the following steps: (1) performing exploratory analysis and outlier detection, (2) testing the basic assumptions (stationarity, homogeneity and serial independence), (3) modeling and estimating model parameters, and (4) making inference, including risk evaluation. In the univariate HFA framework, all these steps are extensively studied and usually considered in the analysis.

Environmental and hydro-meteorological processes, such as floods, droughts, rainstorms, hurricanes, tornadoes, windstorms, weather extremes, and tides, are generally complex. They are often described by more than one correlated variable (e.g., flood volume, peak, and duration), which involves simultaneous consideration of these variables using multivariate models and methods (see e.g., Barnett, 2012). In particular, dealing with extreme hydro-meteorological events requires multivariate HFA. Traditional multivariate HFA methods are too restrictive and do not even apply in many cases. Consequently, over the last two decades, copula functions have emerged as a preferred method in a variety of applications, especially hydro-meteorology and in multivariate HFA.

Multivariate Frequency Analysis of Hydro-Meteorological Variables
DOI: https://doi.org/10.1016/B978-0-323-95908-7.00009-8

Adopting the multivariate HFA framework in hydro-meteorology in preference to univariate HFA was extensively justified in the literature. Indeed, univariate HFA can only provide limited representativity and understanding of extreme events and their probability of occurrence. In addition, the univariate framework of each event characteristic separately does not take into account their dependence structure, leading to potentially less accurate risk estimation. Nevertheless, univariate HFA can be useful in some situations such as when only one variable is significant for design purposes or when the dependence between these variables is not significant. However, multivariate HFA is more reliable for modeling hydro-meteorological variables, leads to better risk assessment, and is a more flexible framework (see Chapter 2 for more details).

1.2 Purpose and aims

Multivariate HFA is a very active research topic in statistical hydrometeorology. A relatively large body of literature, dealing with multivariate HFA, is available mostly as journal papers (theoretical developments, case studies, etc.). In general, these papers treat specific aspects such as a hydrological event (e.g., flood, drought), a particular step of the analysis (e.g., modeling, testing, exploratory analysis), or a given statistical approach or technique (e.g., copulas, L-moments). In addition, most of the literature focuses on the modeling step mainly based on the copula function. Fig. 1.1 illustrates some of the aspects of multivariate HFA highlighted in the literature and their importance and the volume of studies focused on this. It shows that the modeling step, especially based on copulas, is most prevalent the literature. Although copula and modeling are important and essential, many other aspects and steps also need to be considered to perform a complete and appropriate analysis. Multivariate HFA associated literature as papers and reports is not easily accessible to practitioners and students. Hence, this has led to an increasing gap between research and practice in this field. Therefore, the desperate need of a reference book where the reader can find all the relevant material covering the different steps and situations of a multivariate HFA in a simplified and accessible presentation, the connections between them as well as a complete overview of all steps of the analysis has been felt.

This book attempts to reduce or eliminate some of the challenges and difficulties faced by practitioners in multivariate HFA. This book compiles all the relevant background material and new developments in one place and also presents this material in a homogeneous and pedagogical way in order to allow students, engineers, practitioners, and researchers to access and use efficiently all the information about this

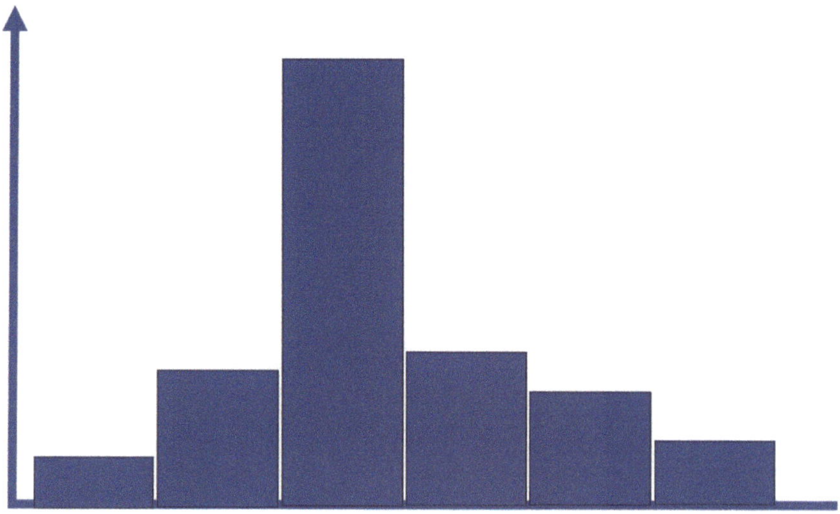

FIGURE 1.1 Illustration of the importance/volume of studies in the literature on each step/topic in multivariate hydrological frequency analysis.

topic. In addition, given the advanced nature of the approaches in multivariate HFA and the ongoing developments, even though useful and necessary, they are complex for a majority of practitioners and students, especially readers without statistical background. Therefore, this book tries to simplify the presentation of these concepts and hence aims to fill the gap between theory and practice. Also, a major part of the literature neglects some of steps of the analysis (Fig. 1.1), potentially leading to incomplete analysis or even wrong conclusions. Consequently, this book highlights the importance of those steps and provides the recent and advanced approaches to deal with them as along with examples from real-life situations.

To the best of the author's knowledge, there is no such existing book that deals specifically and directly with the topic of multivariate HFA as a whole and in an integrated manner. Indeed, the existing books mainly cover copula functions either in hydrology or statistics, such as Salvadori et al. (2007), Zhang and Singh (2019), and Chen and Guo (2019) in water sciences and Joe (2014) and Hofert et al. (2018) in statistics. This book provides a solid platform bringing together multivariate HFA tools in hydrometeorological practice and contributes to filling the gap between theory and practice and the advancement of the field of statistical hydrometeorology. This book enables the reader to perform a well-justified multivariate HFA covering all relevant steps and aspects of the analysis, including the preliminary important steps (e.g., testing the assumptions) and useful extensions (nonstationary, regional). This book provides detailed and comprehensive descriptions of the techniques and all steps involved in performing a complete multivariate HFA.

In this book, the copula-based approach is given due importance and a large chapter (Chapter 5) is dedicated to this topic, along with covering other important topics, including hypothesis testing of the basic assumptions, the return period and quantile, and preliminary analysis such as outliers and descriptive statistics. Where appropriate, some examples based on the same datasets are presented across several chapters to show how to perform the analysis and the steps involved.

1.3 Readership

This book is aimed to be a reference for researchers, practitioners, and graduate students in the field of multivariate HFA, with a clear and comprehensive presentation of all relevant approaches and steps involved in performing a complete analysis. It also serves as an ideal multidisciplinary introductory book for hydrologists, climatologists, and engineers to make themselves familiar with the most up-to-date and advanced multivariate methodologies in hydrological design, planning, and management, to mention some, and their practical applications. This book also serves as a guide for the readers in applying the most recent approaches available toward evaluating hydro-meteorological risks, designing hydraulic structures, and teaching (faculty members), and as state-of-the art methodologies to move rapidly to the next level in their research projects (graduate students and postdocs).

A variety of readers from industry, government agencies, or academia (for research and graduate teaching) as well as statisticians and non-statistician readers can benefit from this book. Advanced approaches are presented in an easy-to-understand manner and with an appropriate level of detail. Even though the primary target readers are hydrologists, climatologists, engineers and statisticians, given that some material is interdisciplinary, it can be used for reference by practitioners from other application fields such as financial institutions, insurance companies (damages caused by floods and droughts), earth sciences, environmental modeling, and government agencies (e.g., public safety, environment and transportation). Readers interested in understanding theoretical concepts and practical aspects related to copula-based modern multivariate HFA can find this book with in-depth technical details extremely helpful, where advanced and complex mathematics/statistics have been avoided to the extent possible. Nevertheless, basic knowledge of probability and statistics, such as random vectors, estimation methods, and statistical tests, is expected.

1.4 Structure and content

Over the last two decades, multivariate HFA has gained icreasing attention in both applications and theoretical developments. This book is composed of seven other chapters and an Appendix. Chapter 2 introduces HFA and briefly makes the connection between the subsequent chapters. Chapters 3–6 discuss the main HFA steps involved in a standard multivariate HFA. Chapters 7 and 8 are dedicated to advanced analysis as well as extensions of the standard analysis, dealing with multivariate nonstationary modeling and multivariate regional analysis. To maintain the fluency of the content, some technical concepts are presented in the Appendix.

Chapter 2 provides an overview and the basics of HFA. It starts with describing the general aims and goals of HFA as well as the essential concepts of return period and quantile for hydrological risk assessment. Then, it briefly introduces the main steps involved in performing the whole HFA. This chapter also discusses the advantages and challenges faced while transitioning from univariate to multivariate HFA frameworks, especially using copula functions. The multivariate character of a number of hydrological phenomena, such as floods, droughts, rainfall storms, and sediments, is described as along with their main features to be treated in the multivariate HFA framework.

Chapter 3 treats the preliminary analysis within the framework of the multivariate HFA. Several statistical properties of the multivariate sample are discussed, such as location, scale, skewness, kurtosis as well as outlier detection. This step can be useful in summarizing, describing, and understanding the information contained in the data series. On the other hand, this preliminary analysis is useful for modeling hydrological variables and hence for risk evaluation. The presented methods are general and can be adapted and applied to a variety of hydro-meteorological events such as floods, droughts, storms, and sediment transport along with other fields.

Chapter 4 addresses the testing step within the framework of multivariate HFA. In this chapter, the corresponding techniques and methods are presented in more detail with a few examples. Nonstationarity, heterogeneity, and serial dependence need to be tested before the modeling step in a multivariate HFA. The testing step is important to ensure the basic assumptions are met and thus the selected models are appropriate. These statistical tests are generic and can be adapted and applied to a variety of hydro-meteorological variables as well as to other fields.

Chapter 5 introduces the modeling step of multivariate HFA based on copula functions. Even though copula modeling is the heart of the multivariate HFA, the preliminary analysis and testing of the basic assumptions should be performed first (presented, respectively, in

Chapters 3 and 4). Here, the basic assumptions are assumed to be fulfilled (see Chapter 4). This chapter also presents an overview of the statistical approaches and methods regarding copula modeling, including parameter estimation and goodness-of-fit testing as well as model selection criteria with illustrations.

Chapter 6 examines the last step of multivariate HFA, which deals with the inference, in particular risk assessment in terms of return periods or quantiles. This step is performed based on the analysis and decisions made in all previous steps (described in previous chapters), especially the modeling step (Chapter 5). Here, risk assessment in hydrology is briefly presented, followed by the basics regarding multivariate return period and quantile, and, finally, an overview of the statistical approaches and methods regarding the selection of the multivariate combinations for a given return period with illustrative examples.

Chapter 7 treats nonstationarity in the multivariate setting. Combining those two aspects (nonstationarity and multivariate) leads to the multivariate nonstationary HFA, which aims at estimating hydro-meteorological quantiles (risks) in the presence of nonstationarity (caused, for instance, by climate change). Prior to performing nonstationary analysis, appropriate tests should be accomplished (Chapter 4). Hydro-climatology phenomena are naturally multivariate with stationarity assumption either fulfilled or not. Therefore, it is more realistic and representative to consider the joint multivariate and nonstationary HFA setting. This chapter briefly introduces the basics of nonstationarity in HFA followed by presenting the multivariate nonstationary context. Then, the modeling methodology of the latter is described followed by an illustrative example.

The last chapter briefly introduces the basics of regional HFA followed by presenting the multivariate context of regional frequency analysis (RFA). Then, the delineation and the regional estimation, as the two main components of RFA, are presented. This chapter also deals with RFA in the multivariate setting. Combining regional and multivariate aspects leads to the multivariate RFA, which aims at estimating hydro-meteorological quantiles (risks) at *ungauged* sites. Usually, in the latter, no hydrological data are available unlike the at-site (local) HFA analysis seen in the previous chapters. RFA in the univariate setting is widely used by hydrologists. The multivariate nature of hydro-meteorological phenomena is present at sites either gauged or ungauged. Therefore, it is more realistic/useful to consider multivariate RFA.

In order to facilitate readability, some of the technical statistical concepts and tools needed in the previous chapters are outlined in the Appendix, which includes ties in the multivariate framework and in hydrology, statistical depth functions, multivariate L-moments, and p-value computation. These tools are also generic and can be useful in other fields and disciplines as well.

1.5 How to read this book?

On one hand, chapters can be followed independently based on the requirements of the reader. For instance, if the objective is to study trends, the reader can directly go to Chapter 4 or if the reader is interested in multivariate modeling, including model selection and parameter estimation, then the reader can find the appropriate material in Chapter 5. On the other hand, if the reader aims to perform a complete multivariate HFA, it is recommended to read the chapters in a sequential order starting from Chapter 2 to Chapter 6 for a standard HFA. Since Chapters 7 and 8 present advanced material, it is recommended to start with the chapters dealing with the standard analysis (Chapters 2–6). This is illustrated in Fig. 1.2, which provides links between all chapters as a diagram showing how they are connected and reading path options.

1.6 Final points

Even though the book attempts to address most of the practical issues and methodological facets related to multivariate HFA, it does not pretend to cover all works in the field. In particular, the applications and case studies are focused on floods even though the presented material and approaches are valid for other hydro-meteorological applications or in other fields as well. Indeed, floods are the most common natural hazards that account for close to half of the total worldwide disasters, affecting hundreds of millions of persons (NDRR, 2020). In addition, multivariate HFA is largely used to study floods. As regards theory, focus was made on the bivariate setting while the high-dimension situations are discussed briefly. The bivariate setting is opted for since it is the most studied case and simple as well. Some other topics like the uncertainty in the estimation as well as vine copulas are only mentionned mainly because this book is a first of its kind in this field, and most of the these concepts are still in development in hydrometeorology.

Even though reasonable efforts have been made to publish reliable data, illustrations, and information, the author and publisher are not responsible for the validity of all the presented materials or the consequences of their use.

The book is entirely based on the subject knowledge of the author. Yet, as in any other publication, it may contain some errors. Comments, corrections, and suggestions from the readers about the material presented in the book and related matters are welcome.

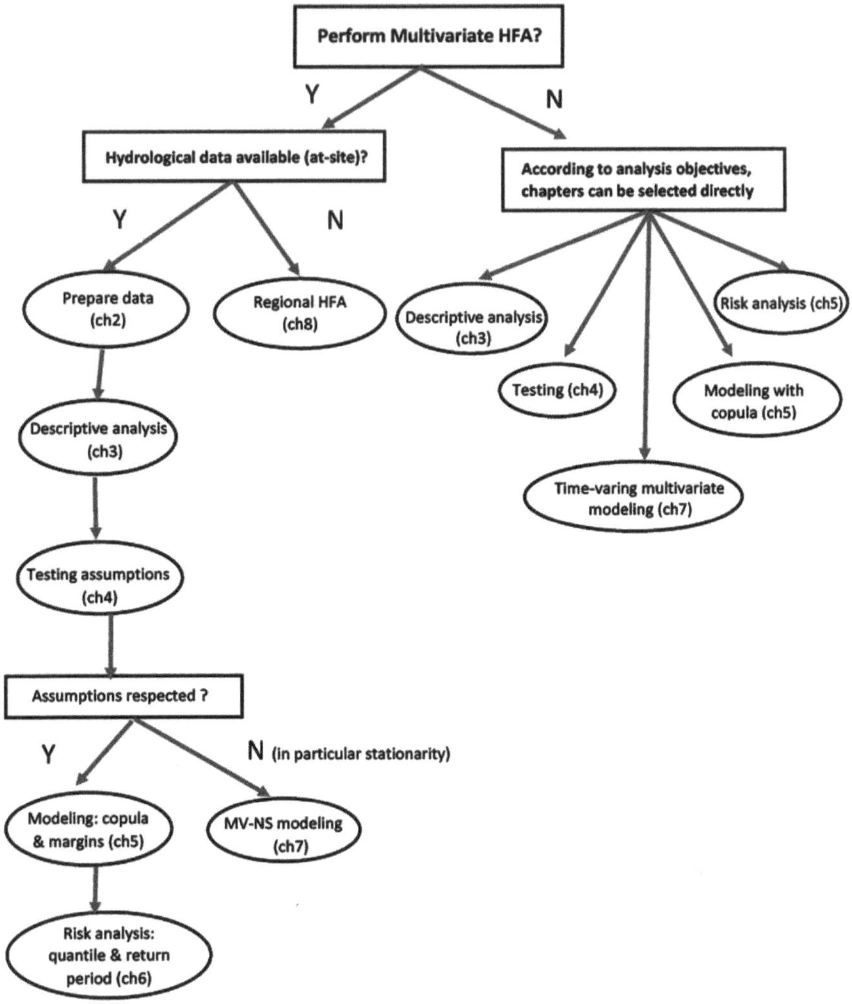

FIGURE 1.2 Illustration of the different path options that can be taken by the reader.

References

Barnett, V. (2012). Outliers, multivariate. *Encyclopedia of Environmetrics*. https://doi.org/10.1002/9780470057339.vao020.

Chen, L., & Guo, S. (2019). *Copulas and its application in hydrology and water resources* (p. 290) Singapore: Springer.

Hofert, M., Kojadinovic, I., Mächler, M., & Yan, J. (2018). *Elements of copula modeling with R* (pp. 9–79). Springer.

Joe, H. (2014). *Dependence modeling with copulas*. CRC Press.

NDRR. (2020). Human cost of disasters: An overview of the last 20 years (2000–2019). https://www.undrr.org/publication/human-cost-disasters-overview-last-20-years-2000-2019.

Salvadori, G., De Michele, C., Kottegoda, N. T., & Rosso, R. (2007). *Extremes in nature: An approach using copulas*. Springer.

Zhang, L., & Singh, V. P. (2019). *Copulas and their applications in water resources engineering*. Cambridge University Press.

CHAPTER

2

Multivariate hydrological frequency analysis, overview

In this chapter, we present an overview and the basics of hydrological frequency analysis (HFA). We start first by describing the general aims and goals of HFA as well as the essential notions of return period and quantile for hydrological risk assessment. Then, we briefly introduce the main steps to perform the whole HFA and refer to the corresponding chapter where each step is fully presented. We cover the advantages and challenges when passing from univariate to multivariate HFA frameworks, especially using copula function. We describe the multivariate character of a number of hydrological phenomena, such as droughts, rainfall storms, and sediments, as well as their main features to be treated in the multivariate HFA framework.

2.1 General aims of hydrological frequency analysis

Serious economic and social consequences may be caused by extreme hydrological events, such as floods, droughts, and storms. Several studies focus specifically on each of these events with different methodologies and for different regions. For instance, in water resources management, droughts is one of the challenging topics whereas extreme rainfall and flood estimates represent the basis for the design of hydraulic and hydrologic structures. A good knowledge concerning the features associated with these extreme hydrological events is important for accurate estimation of the risk associated with water infrastructures in their design and operation. Indeed, on one hand, an underestimation of design floods leads to material damages and loss of human lives. On the other hand, an overestimation leads to an over-sizing of hydraulic structures involving supplementary costs. Hence, it is crucial to consider the appropriate models for the most accurate prediction of these events. To this end,

Multivariate Frequency Analysis of Hydro-Meteorological Variables
DOI: https://doi.org/10.1016/B978-0-323-95908-7.00003-7

HFA, as an ensemble of statistical methods and techniques, is the most considered approach by hydrologists and civil engineers.

The occurrence of extreme events in hydrology cannot be estimated, predicted, or forecasted on the basis of deterministic information with sufficient skill and lead time (e.g., Rao & Hamed, 2000). Alternatively, a probabilistic approach is required to incorporate the effects of such events into decisions. Based on the assumption that occurrences are independent of time, that is, successive events have no relation in their timing and magnitude, HFA can be employed as a decision tool by assessing the likelihood of an event or a combination of events. A number of engineering applications make use of HFA such as hydraulic and municipal structure design (culverts, storm sewers) and landslide hazard evaluation.

Formally, for a random variable X, the main objective of HFA of extreme events is the assessment of the probability $\Pr(X \geq x_T)$ of an event x_T to be exceeded. In statistics, the event x_T is the quantile of probability of exceedance p with $p = \Pr(X \geq x_T)$, whereas in HFA practice, x_T corresponds to the return period T in years. Both notions, quantile and return period, are equivalent since we have

$$\Pr(X \leq x_T) = F\left(x_T; \underline{\theta}\right) = 1 - 1/T, \quad \text{and} \quad x_T = F^{-1}\left(1 - 1/T; \underline{\theta}\right) \tag{2.1}$$

where $\underline{\theta}$ is the vector of parameters associated to the distribution function F. Note that p corresponds to the hydrological risk and x_T is the quantile of order $1 - p$ of the distribution F. Because of the variability of time arrivals between events, a return period is defined as the average of the interevent times between hydrological events. Large return periods are naturally associated to large events and vice versa (Rao & Hamed, 2000).

From Eq. (2.1), fitting a probability distribution F to a series of observed values, such as maximum annual flow, is essential to perform an HFA. Several distributions have been proposed to fit hydrometeorological variables and are commonly used in HFA such as the Generalized Extreme Value (GEV), Lognormal (LN), and the Pearson type 3 family. Specific distributions are recommended in some countries, such as the GEV in the United Kingdom for floods and in the United States for precipitation; the LN in China; and the Log-Pearson type 3 distribution (LP3) in the United States for annual peak flow data. However, the use of the GEV distribution is largely considered because of the simplicity of its quantile function, theoretical results and the availability of software for parameter estimation (El Adlouni et al., 2010). In (Singh, 2017, Chapter 21), a large number of (univariate) distributions are presented and discussed in a hydrological context.

The right tail portion is the most important part of these distributions. Indeed, the hydrological extremes occur for high values of the variable and with very low frequencies. Therefore, El Adlouni et al. (2008)

presented an overview regarding the tail behavior of the main probability distribution families used in HFA. The authors provided a classification of a number of distributions used in HFA according to their tails since a number of distributions could be very similar in their central parts.

Before carrying out a standard HFA, the data series must be independent and identically distributed (iid) (e.g., Rao & Hamed, 2000). In other words, the data series should meet statistical conditions including independence, homogeneity, and stationarity. In a brief and simple way, independence means that the data series do not present any significant autocorrelation. When all the elements of the data series originate from a single population, then the homogeneity is respected. To meet stationarity, except for fluctuations, the data series should be invariant with respect to time and do not have patterns such as trends, jumps, and cycles. In HFA applications, the latter can occur in a variety of situation where for instance jumps are due to a change in station location, whereas long-term climatic fluctuations may lead to trends and cycles. For each of the three assumptions, appropriate statistical tests are usually employed: in the univariate setting, the Wald–Wolfowitz test for independence, the Mann–Whitney test for homogeneity, and the Mann–Kendall for stationarity.

Based on all the above discussed elements, the main and basic steps involved to perform an HFA are as follows: (1) preliminary analysis including an explanatory analysis for data checking and outlier detection; (2) testing that the data series satisfies the hypotheses of independence, homogeneity, and stationarity; (3) fitting of the best theoretical probability distribution to represent the data series; and (4) estimating the quantiles and return period events using the selected distribution. These steps are described in more detail in Section 2.3.

In Eq. (2.1), the parameter vector $\underline{\theta}$ should be estimated. To this end, a number of methods have been developed and are available in the hydrological and statistical literature. The method of moments is among the simplest and most direct ones, which provides parameter estimates such that the theoretical moments are equal to the computed sample moments. Another method for estimating parameters is based on the sample L-moments which is widely used in hydrology, especially in the estimation of GEV distribution parameters. Indeed, sample L-moments are less biased than traditional moment estimators, and thus are better suited for use with small sample sizes commonly encountered in HFA. The method of maximum likelihood (ML) provides estimators that maximize the likelihood function with very interesting statistical properties for large samples which is usually not the case in HFA. In the particular case of the GEV, a generalized ML version can be considered where a prior distribution for the shape parameter of the GEV is involved.

To perform a complete and structured multivariate HFA, the statistical tools, techniques, methods, and models presented in this book should be taken as a whole, respecting their sequential order. However, if taken separately, these statistical tools are also useful to other hydrometeorological and water resources applications, such as simulation, forecasting, downscaling, and weather generation. They are also, in a sense, useful for other boarder applications such as finance and insurance.

2.2 From univariate to multivariate hydrological frequency analysis

In the previous section, we mostly discussed the basic case in HFA where only *one* feature (variable or characteristic) of the hydrological event is involved. It is the univariate HFA. However, generally, hydrological events are characterized by several correlated features, such as peak, volume, and duration for floods; severity, magnitude, and duration for droughts; and storm duration and intensity. Indeed, river management may strongly depend upon the joint features of flood peak and flood volume; the characterization of droughts requires the joint analysis of duration–magnitude–intensity; and different combinations of rainfall intensity and storm duration may generate different storms. Therefore, it is often of fundamental importance to be able to link the marginal distributions of different variables in order to obtain a joint law describing the main features of the hydrological events. The aim of the multivariate HFA framework is to treating the features of each of these events jointly. The multivariate setting is also of interest to study a given feature, for example, flood peak, for two or more different locations (multisite). As a specific example, El Adlouni and Ouarda (2008) estimated the combined risk associated with the flow in the Chateauguay River and the St-Louis Lake in Quebec, Canada. Fig. 2.1 presents an illustration of the variables associated with floods, droughts, and storms, as well as the multisite case.

Multivariate HFA has recently attracted increasing attention where several studies pointed out the importance of considering different variables that characterize a hydrological event (see, e.g., Genest & Chebana, 2017; Hao & Singh, 2016). Several other studies focused on the joint treatment of these event variables based on multivariate techniques such as multivariate distributions or copulas (see, e.g., Chebana & Ouarda, 2011a; De Michele & Salvadori, 2003; Sharma & Mujumdar, 2019; Yue et al., 1999).

In recent years, several studies and review papers were published covering multivariate applications related to extreme events in hydroclimatology (Chebana, 2013; Genest & Chebana, 2017; Hao & Singh, 2016). It was also shown in the literature that univariate HFA provides a limited assessment of an extreme event probability of occurrence. Often, a multivariate analysis is necessary in order to avoid under/overestimation of the risk, as shown below. If a given hydrological event is multivariate,

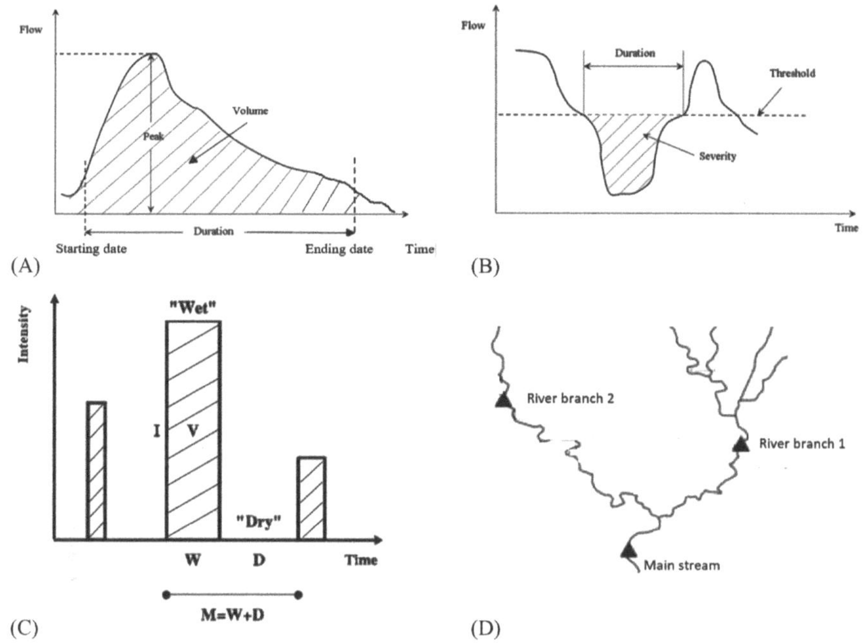

FIGURE 2.1 Different hydrological features in the multivariate setting for the considered events: (A) floods, (B) droughts, (C) rainfall storms, and (D) multisite for flood peak. Source: *(C) Adapted from (Salvadori & De Michele, 2006). Elsevier*

then univariate HFA is not able to provide complete assessment of the probability of occurrence which reduces the accuracy of the risk estimation. The univariate HFA ignores the dependence structure between the event features and hence it is less representative of the phenomenon. However, univariate analysis is useful in situations where only one variable is significant in the design process, or when the random variables have a negligible dependence. The univariate HFA can be considered as a step to be included in the multivariate HFA.

In hydrology, the multivariate framework goes back before 2000 (e.g., Ashkar et al., 1998) where a number of multivariate distributions such as multivariate versions of Normal, Gamma, GEV, and Exponential were developed or employed. Traditionally, the multivariate distributions used in hydrology have normal, log-normal, or exponential margins (Salvadori & De Michele, 2004). However, multivariate distributions are not enough flexible and have a number of limitations. Indeed, available classical multivariate distributions require that the margins to be within the same class and are less flexible regarding the dependence structure. In addition, the usual distributions which can be extended to the multivariate setting are very limited, including for instance normal and gamma (e.g., Hao &

Singh, 2016). These limitations are more problematic for hydrological applications since the event features are usually of very different nature and they are usually related to extreme events.

A construction of multivariate distributions that does not suffer from the drawbacks of classical multivariate models is based on the notion of copulas (Nelsen, 2006). They have been introduced in hydrology in early 2000. A copula is a very useful function to implement efficient algorithms for simulating joint distributions in a more realistic way. In fact, copulas are able to model the dependence structure regardless of the marginal distributions. It is then possible to build multivariate distributions with different margins, the structure of dependence being mathematically formalized through the copula (see Chapter 5 for more details).

The adoption of multivariate framework in hydrology to treat extreme events, versus univariate treatment, was justified in several studies. Salvadori and De Michele (2007) provided the example of a river with two branches where we are interested in the flood peak of both (sum) like Q (branche1) + Q(branche2). They showed indirect applications of copulas as well as the necessity of applying copula and consider the multivariate context where the joint study of the event characteristics leads to a better understanding of the hydrological phenomenon. Therefore, multivariate studies contribute to the improvement of the estimation accuracy and provide information about the dependence structure between hydrological event characteristics (e.g., Hao & Singh, 2016).

Some of the issues that have been addressed, in the multivariate HFA literature, can be summarized as follows: (1) showing the usefulness and the importance of the multivariate framework; (2) selecting the appropriate joint distribution including its copula and marginal distributions, and estimating the associated parameters; (3) introducing and studying the concepts of multivariate return periods and multivariate quantiles; and (4) considering a number of different applications. In the multivariate framework, the matter is not only about the joint distribution although very important, but also about other concepts such as return periods, conditional return period, quantile, and checking the basic assumptions.

When passing from univariate to multivariate HFA, there are some advantages, as discussed before, mainly the dependence between variables, being more realistic, and provides more flexibility to engineers. However, there are some challenges and issues such as variable selection for a given study, consequently dimension selection, as well as event selection. For instance, for floods, one can wonder which variable to include in the analysis among flood peak, volume, and duration. If we choose peak and volume, the dimension would be 2. In the multivariate context, we are interested to study, for instance, probabilities like $\Pr\{X \leq x, Y \leq y\}$ for droughts, and $\Pr\{X \geq x, Y \geq y\}$ when considering floods, where X and Y are one of the above associated variables. Unlike the univariate framework, in the multivariate framework, we find a variety of definitions of each statistical concept, such as for median, quantile, symmetry, and return period (as it will be seen for instance in Chapters 3 and 6). In the multivariate HFA, large data series are required since there are more parameters to estimate, among other reasons.

2.2.1 Main steps of a complete multivariate hydrological frequency analysis

To reach the aims of HFA (either univariate or multivariate), the analysis is mainly composed of the following steps: (1) carrying out an exploratory analysis; (2) checking the basic assumptions involving stationarity, homogeneity and independence; (3) fitting the appropriate model estimating the associated parameters; and (4) making the inference for risk assessment. In the univariate context, all these steps are extensively developed where they are treated in a number of textbooks (e.g., Rao & Hamed, 2000). However, in the multivariate context, the last two steps attract considerably more attention than the first ones. An overview is given in Table 2.1 of the literature in each framework and each step with more focus on the multivariate framework.

TABLE 2.1 Overview of the literature related the main hydrological frequency analysis (HFA) steps in both univariate and multivariate frameworks.

HFA steps	Framework Univariate	Multivariate
Exploratory analysis	*Large literature:* Rao and Hamed (2000)	*Very sparse literature:* Chebana and Ouarda (2011b) Ben Aissia et al. (2017) for missing data
Checking the basic HFA assumptions: • Stationarity • Homogeneity • Independence	*Large literature:* Khaliq et al. (2006)	*Very sparse literature:* Chebana et al. (2010b) for stationarity and homogeneity Chebana et al. (2013) for trend testing Chebana et al. (2017) for shift testing
Modeling and estimation	*Large literature:* Rao and Hamed (2000)	*Large recent literature:* Zhang and Singh (2006) Grimaldi et al. (2016) De Michele and Salvadori (2003)
Risk evaluation and analysis	*Large literature:* Rao and Hamed (2000)	*Limited but growing literature:* Chebana and Ouarda (2011a) Serinaldi (2015) Gräler et al. (2013)

Note: In the univariate framework, some steps are simple and are generally not treated separately. The references are given only as examples from the literature.

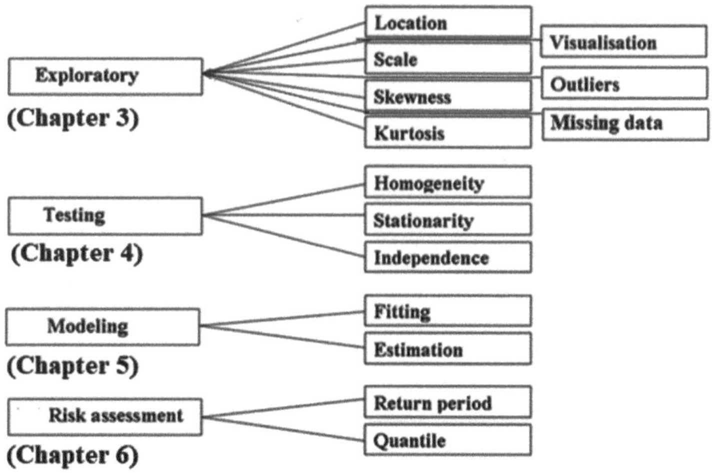

FIGURE 2.2 Overview of the main hydrological frequency analysis steps and their main components as well as the corresponding chapters where they are presented.

Fig. 2.2 illustrates the associated components of each step. These steps can be summarized as follows and a corresponding chapter is presented in the book with more details:

2.2.1.1 *Exploratory analysis*

The sample and distribution features, such as location and scale, but in particular skewness and kurtosis, are of interest for HFA. Indeed, this is because, generally, datasets in HFA are not symmetric and the tail part of the distribution is not negligible (e.g., Hosking & Wallis, 1997). Exploratory analysis is a useful step in order, for instance, to characterize the sample, to assess the similarity between the sample and a known distribution, and to guide the preliminary selection of the distribution. On the other hand, this exploratory analysis allows to detect outliers. The latter can have impacts on the selection of the appropriate distribution and on their estimated parameters (Rao & Hamed, 2000). Chebana and Ouarda (2011b) investigated the features and the shape of bivariate hydrological datasets such as location, scale, skewness, kurtosis, as well as outlier detection. Note that unlike the univariate context, in the multivariate framework, for each sample characteristic (such as median and symmetry) more than one statistical definition is proposed in the literature. This first step is the object of Chapter 3.

2.2.1.2 *Testing basic assumptions*

This step seems to be neglected in the hydrological literature for multivariate datasets. It deals with statistical tests of stationarity, homogeneity, and independence. In the HFA context, Chebana et al. (2010b) reviewed a number of multivariate homogeneity and stationarity testing methods. More specifically, Chebana et al. (2013) reviewed and applied the available multivariate trend tests (related to the stationarity assumption) to floods and multisite peak series. Similarly, Chebana et al. (2017) treated multivariate shift testing (related to the homogeneity assumption) mainly with depth-based tests. Recently, Salvadori et al. (2018) considered risk assessment under multivariate shifts. This step is treated in Chapter 4.

2.2.1.3 *Modeling and parameter estimation*

Compared to all the other steps, this step is the most developed and treated in hydrology and water resources. In order to describe the dependence structure between random variables, copula is the most employed and becomes the standard both in applied and theoretical multivariate modeling. Copulas offer a great flexibility for modeling multivariate samples where they are not dependent on the margins and the latter are not required to be similar or in the same family. Even though copulas have been introduced in 1959, they have recently received increasing attention both in applications and theoretical developments. In the hydrological literature, copula fitting and estimation of the corresponding parameters are among the most studied topics, for example, by Salvatori and De Michele (2004), Kao and Govindaraju (2008), Requena et al. (2016), Zhang and Singh (2006).

Sklar's theorem (Sklar, 1959), as one of the most important result in copula's theory, provides the relationship between a joint multivariate distribution on the one hand and the corresponding copula with the marginal distributions on the other hand. For simplicity, in the bivariate case, Sklar's result shows the existence of a copula C such that:

$$F_{X,Y}(x,y) = C\big(F_X(x), F_Y(y)\big) \quad \text{for all real } x \text{ and } y \tag{2.2}$$

where $F_{X,Y}$ is the joint bivariate distribution of X and Y, and F_X and F_Y are their marginal distributions, respectively. The copula C is unique if F_X and F_Y are continuous, which is common in hydrology. Archimedean and Extreme Value copulas are among the most studied and employed classes in statistics as well as in hydrology. As a typical example of copulas in HFA, we have the Gumbel logistic copula since it is at the same time Archimedean and Extreme Value. For a list of the available copula expressions, the reader is referred to Salvadori et al. (2007) and Zhang and Singh (2019).

In order to select the appropriate copula for a given multivariate dataset, goodness-of-fit tests are more than useful. This is a topic of recent development where for instance, Genest et al. (2009) reviewed and compared several goodness-of-fit tests. In addition, in order to rank or discriminate several accepted copulas based on the above tests, appropriate version of the Akaike information criterion (AIC) criteria should be considered. Note that the commonly employed copulas in hydrometeorology are of one parameter. The latter can be estimated using different methods including the ML method, the maximum pseudo-likelihood (MPL), or the method of moments. The MPL is generally the appropriate method for this purpose. These estimation methods and tests are presented, for instance, in Genest and Nešlehová (2013a,b). The appropriate margins F_X and F_Y are treated in the univariate setting where a number of well-known distribution can be considered such Gumbel, GEV, and Log-Normal (see Rao & Hamed, 2000; Singh, 2017 Chapter 21). For a recent overview of copulas and multivariate modeling in hydrology and water resources, the reader is referred to Genest and Chebana (2017), Hao and Singh (2016), Zhang and Singh (2019). Chapter 5 is dedicated to this important step.

2.2.1.4 Multivariate quantile and return period

In the design of water supply systems and hydraulic structures, the concepts of quantile and return period for hydrometeorological extreme events are commonly used. Briefly, the return period of a given event is defined as the average of the probability of its occurrence. The quantile is the value (s) of the variable(s) corresponding to a given return period. These notions are extended to the multivariate setting. For the multivariate return period, one can refer, for instance, to Gräler et al. (2013), Salvadori and De Michele (2010), and Serinaldi (2015). However, the multivariate quantile was investigated in Chebana and Ouarda (2011a) in the HFA framework. As an example, for the event $\{X>x, Y>y\}$ and annual series, the bivariate return period is defined as the positive number $\hat{T}_{x,y}$ given by:

$$\hat{T}_{x,y} = \frac{1}{\Pr\{X>x, Y>y\}} \tag{2.3}$$

This expression can be adapted for other events, such as $\{X>x \text{ OR } Y>y\}$ and $\{X<x, Y<y\}$, as well as for partial duration series. Relationships between univariate and joint return periods are also derived. The reader is referred to the previous references and to Chapter 6 for more details.

In the statistical literature, a number of multivariate quantile extensions are available. The version employed in hydrology is based on the value of x and y such that $F(x,y) = \Pr\{X \le x, Y \le y\} = p$ for a given p in $]0, 1[$. This version is presented in Belzunce et al. (2007) and adapted to the hydrological context in Chebana and Ouarda (2011a). Since there are an infinity of combinations (x,y) for which $F(x,y) = p$, the corresponding quantile is a curve (not a number as in the univariate case). More precisely, when considering for instance the event $\{X \le x, Y \le y\}$, the quantile curve is given by:

$$Q_{X,Y}(p) = \left\{(x,y) \in R^2 \text{ such that } x = F_X^{-1}(u), y = F_Y^{-1}(v); u, v \in [0,1]: C(u,v) = p\right\} \tag{2.4}$$

It is important to note that this quantile version has some advantages useful for HFA since it is simple, intuitive, interpretable, and probability-based (rather than analytic, algebraic or geometric-based). More descriptions and properties about multivariate quantiles in hydrology can be found in Chebana and Ouarda (2011a) and in Chapter 6 in this book.

2.3 Hydrological events and their main features

In this book, the examples and illustrations are mainly related to floods but other hydrometeorological events such as droughts, rainfall storms and sediments can be considered. These events are related to surface water quantity. They are connected in several ways. For instance, rainfall storms, according to soil saturation, are the origin of floods, which in turn could cause sediment transport. Droughts and low flows are the opposite of floods in terms of water quantity.

A simple cause of floods is a great surplus of water which can result from extreme rainfalls, the rapid thawing of a large accumulations of snow, or a combination of both, depending on location and season. Droughts, with a simple cause as great dearth, may also be considered by analyzing low river flows, but in many areas, droughts result in dry river beds. However, the study of rainfall deficiencies can give more generally meaningful measures of drought, especially to agriculturalists (e.g., Shaw et al., 2010).

Description of hydrological events, their impacts and causes are largely described in many textbooks, general or specific to each event (e.g., Singh, 2017 Chapters 74, 75, 79, and 80). In order to avoid repetition, the focus here is on their multivariate aspects and perspectives. Hence, we will bring from recent papers the multivariate elements for the considered events including their features of interest, their dependence, some applications, models, regions where they were applied.

As previously indicated, each hydrological event can be described by a number of features. These features are generally dependent/correlated and are extracted from the basic crude data, such as daily streamflow series. The multivariate HFA is based and performed on these extracted features. For a given hydrological event, variables/features could have different names or different definitions and ways to extract from daily crude data (see references in Table 2.2). The extraction is a very important step since it affects all the subsequent steps of the analysis, results, and decisions.

2.3.1 Flood features

A flood consists in high water levels overtopping natural, or artificial, banks of a stream, or a river. In most societies, a high price is paid to reduce

TABLE 2.2 Events and their features with examples of selected studies.

	Main features	**Selected references with the studied regions**
Floods	Peak, volume, duration	Ben Aissia et al. (2012), Yue et al. (2001) in Canada Grimaldi and Serinaldi (2006) in Italy and different states in the United States Requena et al. (2016) in Spain
Droughts	Severity, duration, magnitude	Hamdi et al. (2016) in Tunisia Shiau et al. (2007) in China Lee et al. (2013) in Canada and Iran Song and Singh (2010a); Song and Singh (2010b) in the United States
Rainfall storms	Duration, depth, intensity	Vandenberghe et al. (2010) in Belgium Kao and Govindaraju (2007a) in the United States De Michele and Salvadori (2003) in Italy
Sediments	Peak discharge, hydrograph volume, suspended sediment concentration	Bezak et al. (2014), Bezak et al. (2017) in Slovenia and in the United States

the possibilities of damages arising from future floods (e.g., Salvadori et al., 2007). Flood data are collected through streamflow measurements at a relatively sparse network of gages on the main rivers. These data form the primary basis for flood risk analysis and for the design of hydraulic structures.

A number of papers dealt with the study of floods in the multivariate framework including but not limited to Ben Aissia et al. (2012), Grimaldi and Serinaldi (2006), Li et al. (2017), Requena et al. (2016), Salvadori et al. (2018), Zhang and Singh (2007). A flood is described by several features throughout a hydrograph. Fig. 2.1 illustrates a typical hydrograph and the different features. In the following, we give the definition of each flood characteristic as well as the corresponding determination methods from daily flow series. These definitions remain subjective and not unique, especially starting and ending dates.

Starting and ending dates: Starting date ds and ending date de of flood are the most important characteristics since they affect the determination of the remaining features. To calculate these two features, the method proposed by Pacher (2006) is considered and used in a number of studies (e.g., Ben Aissia et al., 2012). It is based on the analysis of cumulative annual hydrographs by adjusting the slopes with a linear approximation.

Flood duration: Once the starting and ending dates are obtained, the duration D is defined by the number of days between the starting date and ending date, that is, $D = de - ds$.

Flood peak: The determination of flood peak Q_p can be obtained (in the simplest way) as the maximum of the daily flow series between *ds* and *de* as $Q_p = \max\{q_i, i = d_s, .., d_e\}$ where q_i is the flow of the day d_i. In order to reduce the 1-day flood peak uncertainty, the flood peak can be defined as the maximum of the *n* flood peak averages around each day *i*. In practice, it is reasonable to take small values of *n* such as $n = 3$ or 7 days. The corresponding flood peak is then $Q_p = \max\{q_{i3}, i = d_s, .., d_e\}$, where q_{i3} is the flow of the day d_i associated to 3 days given by $q_{i3} = \text{mean}\{q_{i-1}, q_i, q_{i+1}\}$.

Flood volume: The volume *V* is determined as $V = \sum_{i=d_s}^{d_e} q_i$, where q_i is the flow of the day d_i.

The above features (D, Q_p, and V) are the main ones and are the most studied in multivariate flood HFA. However, other hydrograph features can be extracted and they attract less attention, especially, in modeling and inference, such as peak, hydrograph shape, and climb-rate (e.g., Ben Aissia et al., 2012).

2.3.2 Illustrative example

In order to illustrate the above flood features, we used daily streamflow data from Nottawasaga River near Baxter (station 02ED003) in Ontario, Canada. Data are available from 1974 to 2012. The obtained flood features are given in Table 2.3. For instance, we observe that the starting date is ranging from February 9 to March 26 whereas the ending date from March 30 and June 3. The corresponding flood durations are from 23 days to 102 days. The flood event is usually caused by snow melting in this river.

2.3.3 Drought features

A drought is the consequence of a climatic fluctuation in which, as commonly conceived, rainfall is unusually low over an extended period and hence the entire precipitation cycle is affected. It may be spread over a year, or longer period, and can affect a large area: a whole country or even a continent. Low flows refer to river flows in the dry period of the year, or the flow of water in a river during prolonged dry weather (e.g., Hamdi et al., 2016; Salvadori et al., 2007). Droughts attracted attention in the multivariate framework where they are studied for instance in De Michele et al. (2013), Hao and AghaKouchak (2014), Hao and Singh (2015), Hamdi et al. (2016), Kim et al. (2003), and Lee et al. (2013).

Given its complexity, a high number of drought definitions are available, such as the one based on the run theory and the drought indices

TABLE 2.3 Extracted main flood features for the Nottawasaga River (station 02ED003).

Year	Starting date (mm-dd)	Ending date (mm-dd)	Duration (days)	Volume (Mm^3)	Peak (m^3/s)	Year	Starting date (mm-dd)	Ending date (mm-dd)	Duration (days)	Volume (Mm^3)	Peak (m^3/s)
1974	02-18	05-12	83	173.68	121.00	1994	02-21	06-03	102	134.55	77.30
1975	03-10	05-13	64	165.60	157.00	1995	03-02	05-15	74	84.55	58.90
1976	03-01	04-30	60	134.01	136.00	1996	03-05	05-15	71	168.77	101.00
1977	02-27	04-14	46	132.81	121.00	1997	02-18	05-03	74	208.35	148.00
1978	03-26	04-30	35	163.15	98.80	1998	02-28	04-20	51	136.42	132.00
1979	03-02	04-30	59	172.97	85.80	1999	02-17	04-28	70	66.31	27.00
1980	03-13	04-30	48	109.33	97.80	2000	02-17	05-07	80	86.11	86.00
1981	02-13	04-27	73	120.63	112.00	2001	02-23	04-29	65	152.30	72.90
1982	03-15	04-19	35	150.30	142.00	2002	02-14	05-21	96	124.80	40.60
1983	03-03	05-25	83	149.00	46.00	2003	03-08	05-10	63	84.75	44.00
1984	02-13	05-05	82	192.08	70.00	2004	02-26	04-27	61	133.24	132.00
1985	02-24	04-27	62	232.45	100.00	2005	03-17	05-08	52	127.93	81.60
1986	03-04	04-22	49	130.27	88.70	2006	02-22	04-28	65	145.56	133.00
1987	03-06	04-19	44	126.34	106.00	2007	03-08	04-27	50	100.07	97.00
1988	02-25	05-04	69	106.59	52.00	2008	03-14	04-29	46	151.47	159.00
1989	03-11	04-25	45	89.52	86.00	2009	02-09	05-13	93	223.45	131.00
1990	02-22	04-22	59	156.75	243.00	2010	03-07	03-30	23	75.44	105.00
1991	02-26	05-06	69	174.50	90.40	2011	02-20	05-21	90	220.79	102.00
1992	03-01	05-10	70	112.01	76.40	2012	02-24	05-13	79	116.20	72.30
1993	03-20	04-29	40	131.27	126.00						

(see, e.g., Zhang et al., 2015). A drought is, however, a multivariate event characterized by its *volume* V_d or *severity* S_d representing the water deficit below a selected threshold T_r, its *duration* D_d and *magnitude* M_d given by the ratio of the severity over the duration, as given by:

$$V_d = T_r D_d - \int_{t_b}^{t_e} q_t dt = \int_{t_b}^{t_e} (T_r - q_t) dt \quad q_t < T_r, \quad M_d = V_d / D_d \tag{2.5}$$

where t_b and t_e are respectively the starting and ending times of the drought event and q_t is the discharge at the time t.

To ensure independence of the events, a drought is defined, using the threshold level method, as an event during which the streamflow is continuously below a certain level as shown in Fig. 2.1. The threshold level method has been evaluated for its applicability to daily discharge series for streams in different climate zones and with different hydrological regimes.

For illustrative example of the extracted drought features, we cite for instance those given in Hamdi et al. (2016). It is about the Medjerda River, the longest and most important river in Tunisia. It constitutes the water supply source for more than half of the Tunisian population. The authors employed daily streamflow data for the Jendouba station from 1966 to 2008. Different thresholds T_r were considered. For instance, for $T_r = 0.387\ m^3/s$ (corresponding to 90% low flow from the daily flow duration curve, FDC), 32 drought events were detected. For a given threshold, the drought features D_d, S_d and M_d were extracted from the observed drought events.

2.3.4 Rainfall storm features

The analysis of rainfall occurrence depends fundamentally on the length of the rainfall duration for which the information is required. In describing the measurement of precipitation, it has been emphasized that most data are provided by daily gauges, but it is the recording gauges (and radar) that identify the shorter time-scale incidence of rain and give measures of rainfall quantities related to time (Shaw et al., 2010).

A number of papers dealt with the study of the rain storm in the multivariate framework (e.g., Balistrocchi & Bacchi, 2011; De Michele & Salvadori, 2003; Kao & Govindaraju, 2007b; Vandenberghe et al., 2010). Salvadori and De Michele (2007) provided an interesting example of copula application in the case of storms.

To characterize rainfall, a useful measure is the average intensity (I_{rain}) that is expressed as rainfall depth (ED_{rain}) divided by its duration (D_{rain}). Then, the rainfall event can be viewed as a rectangular pulse with duration as the width and average rainfall intensity as the magnitude of this pulse.

Interesting illustrative examples can be found in the aforementioned papers and in particular Kao and Govindaraju (2007b). They provided a discussion on event selection (see their Section 2.3) and treated the features depth, duration and intensity. The case study is about data from Indiana, United Sates. Another interesting example can be found in Salvadori and De Michele (2007) using data from Northern Italy.

2.3.5 Sediment features

The sediment transport and silting of reservoirs constitute, by their importance, a major problem in several countries. Major challenges are associated with the issue of sediment transport where the processes are complex, the hydro-climatic context is variable and data is lacking.

In contrast to floods and droughts, HFA attracted less attention in the case of sediments. HFA, in the sediment context, was first used on flow data to determine the "effective discharge," that is, the flow transporting most sediment over a given period, but not directly on sediment data itself. HFA of flow data was also combined to other sediment estimation methods to derive future sediment transport. It is only in the last two decades that HFA procedures were performed directly on sediment variables such as suspended sediment concentration or sediment discharge (Benkhaled et al., 2014; Higgins et al., 2011; Soler et al., 2007; Tramblay et al., 2010; Watts et al., 2003). Multivariate studies on sediment are rare, may be because of data scarcity. Bezak et al. (2014) studied flood features (volume and peak) with suspended sediment loads (SSL) where a copula function is used. SSL are usually correlated with peak discharge values and consequently also with hydrograph volumes, so these hydrological phenomena are multivariate. In this framework, one can find the more recent study by Bezak et al. (2017). Given rare studies in multivariate HFA dealing with sediments, examples of the extracted sediment features can be found in the aforementioned references, that is, Bezak et al. (2014, 2017).

References

Ashkar, F., El Jabi, N., & Issa, M. (1998). A bivariate analysis of the volume and duration of low-flow events. *Stochastic Hydrology and Hydraulics, 12*, 97–116.

Balistrocchi, M., & Bacchi, B. (2011). Modelling the statistical dependence of rainfall event variables by a trivariate copula function. *Hydrology and Earth System Sciences Discussions, 8*, 429–481.

Belzunce, F., Castaño, A., Olvera-Cervantes, A., & Suárez-Llorens, A. (2007). Quantile curves and dependence structure for bivariate distributions. *Computational Statistics & Data Analysis, 51*(10), 5112–5129.

Ben Aissia, M. A., Chebana, F., Ouarda, T. B., Roy, L., Desrochers, G., Chartier, I., & Robchaud, É. (2012). Multivariate analysis of flood characteristics in a climate change context of the watershed of the Baskatong reservoir, Province of Québec, Canada. *Hydrological Processes, 26*(1), 130–142.

Ben Aissia, M. A., Chebana, F., & Ouarda, T. B. (2017). Multivariate missing data in hydrology–Review and applications. *Advances in Water Resources, 110*, 299–309.

Benkhaled, A., Higgins, H., Chebana, F., & Necir, A. (2014). Frequency analysis of annual maximum suspended sediment concentrations in Abiod wadi, Biskra (Algeria). *Hydrological Processes, 28*(12), 3841–3854.

Bezak, N., Mikoš, M., & Šraj, M. (2014). Trivariate frequency analyses of peak discharge, hydrograph volume and suspended sediment concentration data using copulas. *Water Resources Management, 28*(8), 2195–2212.

Bezak, N., Rusjan, S., Kramar Fijavž, M., Mikoš, M., & Šraj, M. (2017). Estimation of suspended sediment loads using copula functions. *Water, 9*(8), 628.

Chebana, F. (2013). *Multivariate analysis of hydrological variables, in encyclopedia of environmetrics.* (2nd ed.). Chichester: Wiley.

Chebana, F., Aissia, M. A. B., & Ouarda, T. B. (2017). Multivariate shift testing for hydrological variables, review, comparison and application. *Journal of Hydrology, 548*, 88–103.

Chebana, F., & Ouarda, T. B. M. J. (2011a). Multivariate quantiles in hydrological frequency analysis. *Environmetrics, 22*(1), 63–78.

Chebana, F., & Ouarda, T. B. M. J. (2011b). Depth-based multivariate descriptive statistics with hydrological applications. *Journal of Geophysical Research, 116*(D10), D10120.

Chebana, F., Ouarda, T. B., & Duong, T. C. (2013). Testing for multivariate trends in hydrologic frequency analysis. *Journal of Hydrology, 486*, 519–530.

Chebana, F., Ouarda, T. B. M. J., & Duong, T. C. (2010b). Testing for stationarity, homogeneity and trend in multivariate hydrologic time series: A review. *Reports*, I-271.

De Michele, C., & Salvadori, G. (2003). A generalized pareto intensity-duration model of storm rainfall exploiting 2-Copulas. *Journal of Geophysical Research D: Atmospheres, 108*(2).

De Michele, C., Salvadori, G., Vezzoli, R., & Pecora, S. (2013). Multivariate assessment of droughts: Frequency analysis and dynamic return period. *Water Resources Research, 49* (10), 6985–6994.

El Adlouni, S., & Ouarda, T. B. M. J. (2008). Study of the joint law flow-level by copulas: Case of the Chateauguay river. *Canadian Journal of Civil Engineering, 35*(10), 1128–1137.

El Adlouni, S., Bobée, B., & Ouarda, T. B. M. J. (2008). On the tails of extreme event distributions in hydrology. *Journal of Hydrology, 355*(1–4), 16–33.

El Adlouni, S., Chebana, F., & Bobée, B. (2010). Generalized extreme value versus Halphen system: Exploratory study. *Journal of Hydrologic Engineering, 15*(2), 79–89.

Genest, C., & Chebana, F. (2017). In V. P. Singh (Ed.), *Handbook of applied hydrology* (2nd ed., pp. 301–309). New York.: McGraw-Hill, Chapter 30.

Genest, C., & Nešlehová, J. G. (2013a). Assessing and modeling asymmetry in bivariate continuous data. *In Copulae in mathematical and quantitative finance* (pp. 91–114). Berlin, Heidelberg: Springer.

Genest, C., & Nešlehová, J. (2013b). Copulas and copula models. *Encyclopedia of Environmetrics.*

Genest, C., Rémillard, B., & Beaudoin, D. (2009). Goodness-of-fit tests for copulas: A review and a power study. *Insurance: Mathematics and Economics, 44*(2), 199–213.

Gräler, B., van den Berg, M., Vandenberghe, S., Petroselli, A., Grimaldi, S., De Baets, B., & Verhoest, N. (2013). Multivariate return periods in hydrology: A critical and practical review focusing on synthetic design hydrograph estimation. *Hydrology and Earth System Sciences, 17*(4), 1281–1296.

Grimaldi, S., & Serinaldi, F. (2006). Asymmetric copula in multivariate flood frequency analysis. *Advances in Water Resources, 29*(8), 1155–1167.

Grimaldi, S., Petroselli, A., Salvadori, G., & De Michele, C. (2016). Catchment compatibility via copulas: A non-parametric study of the dependence structures of hydrological responses. *Advances in Water Resources, 90*, 116–133.

Hamdi, Y., Chebana, F., & Ouarda, T. B. (2016). Bivariate drought frequency analysis in the Medjerda river basin. *Tunisia. Journal of Civil & Environmental Engineering, 6*, 1000227.

Hao, Z., & AghaKouchak, A. (2014). A nonparametric multivariate multi-index drought monitoring framework. *Journal of Hydrometeorology, 15*(1), 89–101.

Hao, Z., & Singh, V. P. (2015). Drought characterization from a multivariate perspective: A review. *Journal of Hydrology, 527*, 668–678.

Hao, Z., & Singh, V. P. (2016). Review of dependence modeling in hydrology and water resources. *Progress in Physical Geography, 40*(4), 549–578.

Higgins, H., St-Hilaire, A., Courtenay, S. C., & Haralampides, K. A. (2011). Suspended sediment dynamics in a tributary of the Saint John River, New Brunswick. *Canadian Journal of Civil Engineering, 38*(2), 221–232.

Hosking, J. R. M., & Wallis, J. R. (1997). *Regional frequency analysis: An approach based on L-moments* (p. 240) Cambridge: Cambridge University Press.

Kao, S.-C., & Govindaraju, R. S. (2007a). Probabilistic structure of storm surface runoff considering the dependence between average intensity and storm duration of rainfall events. *Water Resource Research, 43*, W06410. Available from https://doi.org/10.1029/2006WR005564.

Kao, S.-C., & Govindaraju, R. S. (2007b). A bivariate frequency analysis of extreme rainfall with implications for design. *Journal of Geophysical Research, 112*(D13), D13119.

Kao, S. C., & Govindaraju, R. S. (2008). Trivariate statistical analysis of extreme rainfall events via the Plackett family of copulas. *Water Resources Research, 44*, 2.

Khaliq, M. N., Ouarda, T. B. M. J., Ondo, J. C., Gachon, P., & Bobée, B. (2006). Frequency analysis of a sequence of dependent and/or non-stationary hydro-meteorological observations: A review. *Journal of Hydrology, 329*(3-4), 534–552.

Kim, T. W., Valdés, J. B., & Yoo, C. (2003). Nonparametric approach for estimating return periods of droughts in arid regions. *Journal of Hydrologic Engineering, 8*(5), 237–246.

Lee, T., Modarres, R., & Ouarda, T. B. (2013). Data-based analysis of bivariate copula tail dependence for drought duration and severity. *Hydrological Processes, 27*(10), 1454–1463.

Li, J., Chen, Y., Wang, H., Qin, J., Li, J., & Chiao, S. (2017). Extending flood forecasting lead time in a large watershed by coupling WRF QPF with a distributed hydrological model. *Hydrology and Earth System Sciences, 21*(2), 1279–1294.

Nelsen, R. B. (2006). *An introduction to copula* (2nd ed.). New York: Springer.

Pacher, G. W. (2006). Détermination objective des paramètres des hydrogrammes. *Ouranos*, 8, In French.

Rao, A. R., & Hamed, K. H. (2000). *Flood frequency analysis* (p. 376) Boca Raton, FL: CRC Press.

Requena, A. I., Chebana, F., & Mediero, L. (2016). A complete procedure for multivariate index-flood model application. *Journal of Hydrology, 535*, 559–580.

Salvadori, G., & De Michele, C. (2004). Frequency analysis via copulas: Theoretical aspects and applications to hydrological events. *Water Resources Research, 40*(12).

Salvadori, G., & De Michele, C. (2006). Statistical characterization of temporal structure of storms. *Advances in Water Resources, 29*(6), 827–842.

Salvadori, G., & De Michele, C. (2007). On the use of copulas in hydrology: Theory and practice. *Journal of Hydrologic Engineering, 12*(4), 369–380.

Salvadori, G., & De Michele, C. (2010). Multivariate multiparameter extreme value models and return periods: A copula approach. *Water Resources Research, 46*, 10.

Salvadori, G., De Michele, C., Kottegoda, N. T., & Rosso, R. (2007). *Extremes in nature: An approach using copulas* (56). Springer Science & Business Media.

Salvadori, G., Durante, F., De Michele, C., & Bernardi, M. (2018). Hazard assessment under multivariate distributional change-points: Guidelines and a flood case study. *Water, 10*, 751.

Serinaldi, F. (2015). Dismissing return periods!. *Stochastic Environmental Research and Risk Assessment, 29*(4), 1179–1189.

Sharma, S., & Mujumdar, P. P. (2019). On the relationship of daily rainfall extremes and local mean temperature. *Journal of Hydrology, 572*, 179–191.

Shaw, E. M., Beven, K. J., Chappell, N. A., & Lamb, R. (2010). *Hydrology in practice*. CRC press.

Shiau, J. T., Feng, S., & Nadarajah, S. (2007). Assessment of hydrological droughts for the Yellow River, China, using copulas. *Hydrological Processes, 21*(16), 2157–2163.

Singh, V. P. (Ed.), (2017). *Handbook of applied hydrology*. New York: McGraw-Hill Education.

Sklar, M. (1959). Fonctions de répartition à n dimensions et leurs marges. *Publications de l'Institut de statistique de l'Université de Paris, 8*, 229–231.

Soler, M., Regüés, D., Latron, J., & Gallart, F. (2007). Frequency–magnitude relationships for precipitation, stream flow and sediment load events in a small Mediterranean basin (Vallcebre basin, Eastern Pyrenees). *Catena, 71*(1), 164–171.

Song, S., & Singh, V. P. (2010a). Meta-elliptical copulas for drought frequency analysis of periodic hydrologic data. *Stochastic Environmental Research and Risk Assessment, 24*(3), 425–444.

Song, S., & Singh, V. P. (2010b). Frequency analysis of droughts using the Plackett copula and parameter estimation by genetic algorithm. *Stochastic Environmental Research and Risk Assessment, 24*(5), 783–805.

Tramblay, Y., Saint-Hilaire, A., Ouarda, T. B., Moatar, F., & Hecht, B. (2010). Estimation of local extreme suspended sediment concentrations in California Rivers. *Science of the Total Environment, 408*(19), 4221–4229.

Vandenberghe, S., Verhoest, N. E. C., Buyse, E., & De Baets, B. (2010). A stochastic design rainfall generator based on copulas and mass curves. *Hydrology and Earth System Sciences, 14*(12), 2429–2442.

Watts, C. D., Naden, P. S., Cooper, D. M., & Gannon, B. (2003). Application of a regional procedure to assess the risk to fish from high sediment concentrations. *Science of the Total Environment, 314*, 551–565.

Yue, S., Ouarda, T., & Bobée, B. (2001). A review of bivariate gamma distributions for hydrological application. *Journal of Hydrology, 246*(1-4), 1–18.

Yue, S., Ouarda, T. B. M. J., Bobée, B., Legendre, P., & Bruneau, P. (1999). The Gumbel mixed model for flood frequency analysis. *Journal of Hydrology, 226*(1-2), 88–100.

Zhang, L., & Singh, V. P. (2019). *Copulas and their applications in water resources engineering*. Cambridge University Press.

Zhang, L., & Singh, V. P. (2006). Bivariate flood frequency analysis using the copula method. *Journal of Hydrologic Engineering, 11*(2), 150–164.

Zhang, L., & Singh, V. P. (2007). Gumbel-Hougaard copula for trivariate rainfall frequency analysis. *Journal of Hydrologic Engineering, 12*(4), 409–419.

Zhang, Q., Qi, T., Singh, V. P., Chen, Y. D., & Xiao, M. (2015). Regional frequency analysis of droughts in China: A multivariate perspective. *Water Resources Management, 29*(6), 1767–1787.

CHAPTER

3

Multivariate preliminary analysis

In this chapter, we treat the preliminary analysis within the framework of the multivariate hydrological frequency analysis (HFA). It represents the first step briefly described in Chapter 2. Several statistical properties of the multivariate sample are treated, such as location, scale, skewness, kurtosis, as well as outlier detection. The usefulness of this preliminary step is for its own as well as for further analysis. It can be useful to summarize, describe, and understand the information contained in the data series. On the other hand, it is useful for modeling of hydrological variables and hence for risk evaluation. It allows screening the data, to guide selecting the appropriate model as well as performing comparisons of multivariate samples. These methods are general and can be adapted and applied to a variety of hydrological events such as floods, droughts, storms, and sediment transport. The preliminary analysis should be done correctly but is not necessary to be dominant in the whole study, like in HFA.

3.1 Context and motivation

In the multivariate HFA literature, a number of issues have been addressed and largely studied whereas the preliminary and exploratory analysis has attracted less attention (Chebana, 2013). Among the issues that have been largely studied, we found those related to the inferential statistics, in particular, copula modeling and parameter estimation, as well as multivariate return periods and quantiles. However, any statistical analysis should begin with an investigation and a close inspection of the data. Accordingly, if the data are found to be appropriate, such as in terms of quantity and quality, further analysis can be undertaken. In the univariate HFA setting, this step is usually included in the analysis (e.g., Helsel & Hirsch, 2002), whereas it is almost neglected in the multivariate HFA literature.

Multivariate Frequency Analysis of Hydro-Meteorological Variables
DOI: https://doi.org/10.1016/B978-0-323-95908-7.00001-3

Quantifying and summarizing statistical properties of samples and distributions is one of the main aims of exploratory analysis. Fig. 2.1 in Chapter 2 provides an overview of the composition of this step as part of HFA. Usefulness of exploratory analysis can be seen in different ways. First, it can be helpful to guide the selection of the appropriate distribution based on some graphics and summary statistics. In addition, it can be employed to understand the nature of the phenomenon that generates the data. For instance, centrality, dispersion, symmetry, and peakedness are some of the important statistical properties of a sample. They can be respectively measured by the location, scale, skewness, and kurtosis. Indeed, location and scale are summary statistics, whereas the shape can be captured by skewness and kurtosis. In the multivariate case, (cross)-dependence assessment is an additional and specific element in the analysis via a number of dependence measures, for example, correlation and Kendall's coefficients. Hydrologic data mostly have very skewed, nonnormal distributions and often display heavy tails where the latter is related to risk assessment as well as to kurtosis (Helsel & Hirsch, 2002).

One of the important tasks in an exploratory analysis is related to outlier detection and treatment. In a simple fashion, outliers can be unusual observations, or gross errors and inconsistencies in the dataset. Hence, they can negatively impact subsequent analysis such as the selection of the appropriate distribution and/or the estimation of the corresponding parameters. Consequently, in order to ensure that the inference and decisions are relying on the right data set, detection and management of outliers should be considered (Barnett & Lewis, 1998). In the context of hydrological data, outliers can be present since the data may be incorrect and/or circumstance changes related to the measurements may occur over time (Rao & Hamed, 2000).

For univariate samples and distributions, the above concepts are well defined and their computation is straightforward. Several techniques were directly inspired by univariate ones and developed in an analogous manner leading to the classical multivariate analysis. The corresponding literature is very wide and extensive where a variety of textbooks are available (e.g., Anderson, 1984). A number of these classical multivariate methods (such as clustering or principal component analysis) have different aims than of multivariate HFA. They are mainly component-wise and are based on multivariate normal distribution and moments. Since in the multivariate setting, variables are usually mutually dependent, component-wise techniques perform poorly. On the other hand, moment-based methods depend on the existence of moments. Usually, in HFA, the data are not normally distributed and the components of an event are (cross) dependent. Hence, classical multivariate methods are not appropriate for HFA purposes. The reader is referred to Anderson (1984) for a detailed review of classical multivariate techniques.

Recently developed techniques avoid the above drawbacks of the classical approaches. The former are based on depth functions, which in short are a multivariate inward-outward ranking functions. Depth-based techniques are not component-wise, do not rely on normality, and they are moment-free and affine invariant if the depth function is. These advantages are useful, for instance, to include distributions appropriate for hydrological applications such as Cauchy and other nonnormal distributions. The depth-based ranking also enables numerous outlier detection techniques. Chebana and Ouarda (2011) focused on the application of the depth-based approaches to the exploratory step of multivariate HFA. More description and illustrations of depth functions are given in Appendix.

It is important to indicate that, in the multivariate context, more than one definition can be found for each sample feature, such as centrality and symmetry, which is not the case for the univariate setting. Liu et al. (1999) can be considered as a key reference in the study of multivariate descriptive statistics. It covers most of the above-mentioned features. However, a number of other studies focused on each sample feature separately such as multivariate outlier detection that was studied by Dang and Serfling (2010).

The aim of this chapter is to deal with the first preliminary analysis step within the multivariate HFA context. Let $X_1, X_2, \ldots, X_n \in R^d$ be a d-dimensional ($d \geq 1$) sample with size $n \geq d$. In the following, almost all the presented statistical technics and tools are based on the notion of depth function. Basically, a depth function is a real-valued function that provides a ranking of a multivariate sample. More details on depth functions are given in Appendix. Using a given depth function $D(.)$, the sample can be sorted in decreasing order of depth values to obtain $X_{[1]}, X_{[2]}, \ldots, X_{[n]}$. Then, the *"de-class"* of $X_{[i]}$ is defined as the set of observations having the same depth value, for $i = 1, \ldots, n$.

3.2 Visualization

Before conducting any statistical analysis, data should be visualized. At the least, visualization tools allow to be sure about the data to be analyzed. Important features of the data may be missed or misunderstood solely on the basis of numerical summaries without appropriate graphs. A typical illustrative example of this aspect could be the case where two series have a very low correlation coefficient value whereas their scatter plot clearly shows a relationship (not linear) between the series. On the other hand, in theory as well in practice, statistical methods and models, including multivariate HFA, require a number of assumptions. Visualization techniques are useful, but not alone, for checking these assumptions. A key point of visualizing data is that data "should have the opportunity to speak for themselves, prior to or a part of a formal analysis" (Maindonald & Braun, 2010). Behind the statistical development of graphics and visualization tools, advances in computing make them more practical and facilitate their use as well as improving their capacity. Indeed, data visualization is a very evolving topic and goes beyond the field of statistics, especially in the age of machine learning and data mining where for complex data sets, in terms of high dimensional, spatial, temporal, and large quantity of information, sophisticated techniques could be considered. Note that even with higher dimensional data, we are often interested in looking at two- or three-dimensional projections. Most of the interesting features occur in lower dimensions where the bivariate and trivariate cases are very important and useful, such as in HFA but not the case in other hydrological and water resource applications.

Visualizing data and graphical techniques have a number of advantages including for instance (see Maindonald & Braun, 2010):

- It may suggest ideas and understandings that had not previously been contemplated.
- It may challenge the theoretical understanding that guided the initial collection of the data.
- It allows data to criticize an intended analysis and facilitates checks on assumptions. Subsequent formal analysis can then proceed with greater confidence.
- It may reveal additional information, not directly related to the research question. It may, for example, suggest fruitful new lines of research.

In particular for multivariate data, visualizing data and graphical techniques have been the object of several entire books or chapters. A summary of these techniques can be found in the chapter by McLeod and Provost (2006).

The scatterplot is the simplest visualization tool in the two- or three-dimensional cases. As an extension for higher dimensions, we have the scatterplot matrix which shows the pairwise scatterplots. In addition, the boxplot is one of the most used visualization tools. It is extended to the bivariate case and called the bagplot. The bagplot generally gives clues about the sample, such as location, dispersion, and shape. The bagplot is composed by a central bag, a center, and a fence. The central bag covers half of the deepest points. Regarding the center, it can be taken as the Tukey median, defined below. The fence of the bagplot is a region delimited by the points included in the central bag inflated by a factor 3. Points outside the fence region are considered as potential outliers.

Another bivariate plot is the sunburst plot. It is similar to the bagplot. The former is based either on Tukey or Liu depth functions (given in Appendix), whereas the bagplot is based on Tukey depth function. In addition, the sunburst plot does not have a fence region which makes it unappropriate to detect potential outliers. A formal approach is considered below to detect multivariate outliers.

In order to reveal the shape and structure of a given multivariate dataset, the contours of the depth function can be considered as a part of the visualization of the data. Contours can also be used to compare between bivariate data sets. One of the way they can be evaluated is on the basis of depth functions for which the Tukey depth function is the most used and studied.

Example

In this example, we treat data of the example of Chapter 2 using data from Nottawasaga River near Baxter (station 02ED003) in Ontario, Canada. We treat the flood peak (Q) and volume (V) series. First, we present the corresponding scatter plot in Fig. 3.1.

The bagplot corresponding to this data set, based on Tukey depth, is presented in Fig. 3.2. The orientation of the bag indicates positive dependence between the flood variables Q and V. This is in agreement with a number of studies in the multivariate flood FA literature (e.g., Zhang & Singh, 2006). Note that the points outside the fence in Fig. 3.2 correspond to the flood of 1990. It has the smallest depth value. As indicated previously, it is suspected but cannot be considered as formal outlier at this stage of the analysis.

The contours of (Q, V), in Fig. 3.3, are close to be circular and seem to be distant to each other representing dispersion of the data series.

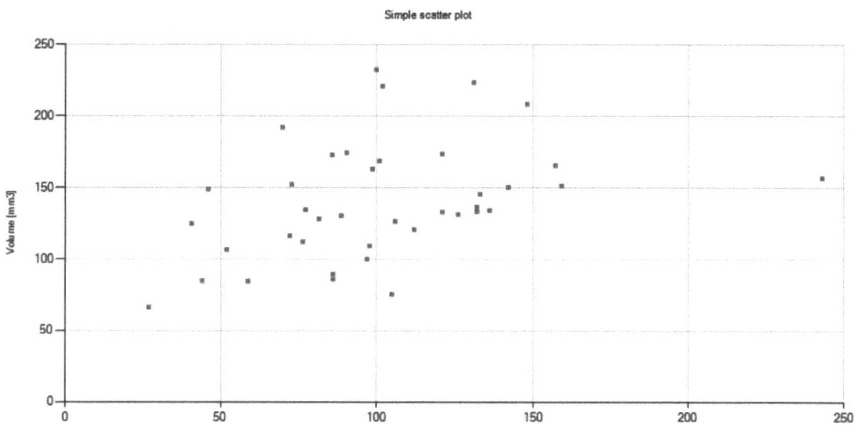

FIGURE 3.1 Scatter plot of (Q, V) associated to Nottawasaga River.

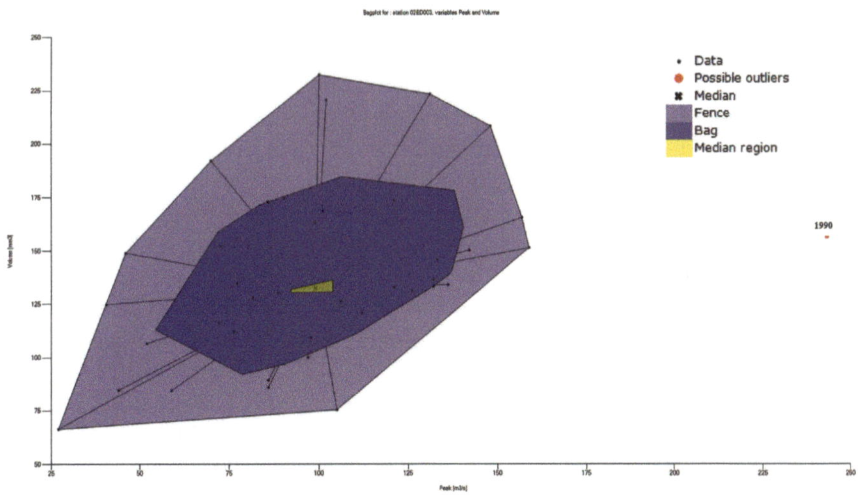

FIGURE 3.2 Bagplot of the (Q,V) series for the Nottawasaga River.

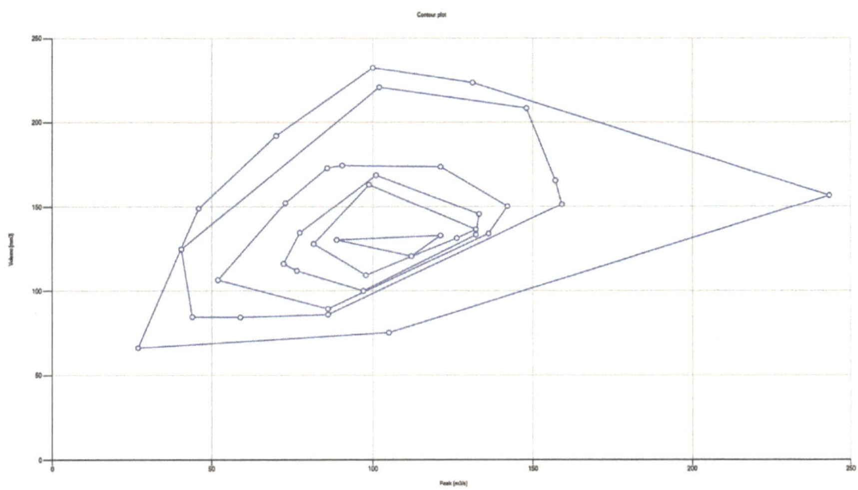

FIGURE 3.3 Contour plot of (Q,V) series for the Nottawasaga River.

3.3 Cross-dependence measures

Within the multivariate framework, where we deal with two or more variables, the first question that comes to mind is about the correlation or the dependence between the variables. Intuitively, a measure of dependence indicates how closely two random variables X and Y are related. However, different types of dependence can occur and are of interest, that is, serial- and cross-dependence as well as overall the variability range of the variables or focusing on some parts such as the extremal parts. The serial dependence deals with multivariate *observations* such as $(X_1, Y_1), \ldots, (X_n, Y_n)$ for two random variables X and Y where $(X_1, \ldots, X_n)$ and $(Y_1, \ldots, Y_n)$ could be, respectively, for instance, peak flow series and flood volume series. The dependence between the component variables X and Y is called cross-dependence.

The term "cross" is specified to distinguish with "serial" dependence where both are considered in the multivariate HFA. In this section, we focus on the first one whereas the second one is treated in Chapter 4 dealing with hypothesis testing. On the other hand, in this section, we deal with overall or central dependence measures and not on specific parts of the distribution such as the tails, which is presented in Chapter 5.

The Pearson's correlation coefficient is considered as the most familiar measure of dependence between two series or variables. It is commonly called "the correlation coefficient." The population correlation coefficient between two random variables X and Y is defined as

$$\rho_P = corr(X, Y) = \frac{\text{cov}(X, Y)}{\sigma_X \sigma_Y} = \frac{E\left[(X - \mu_X)(Y - \mu_Y)\right]}{\sigma_X \sigma_Y} \tag{3.1}$$

where μ_X and μ_Y are the corresponding means, σ_X and σ_Y are the corresponding standard deviations, E is the expected value operator, *cov* is the covariance, and *corr* refers to the correlation.

Kendall's and Spearman's coefficients are other well-known dependence measures. The Kendall's tau (τ) coefficient is given by

$$\tau = \frac{(\text{number of concordant pairs}) - (\text{number of discordant pairs})}{n(n-1)/2} \tag{3.2}$$

where for $i, j = 1, \ldots, q$, and (x_j, y_j) are said to be concordant if $x_i > x_j$ and $y_i > y_j$ or if $x_i < x_j$ and $y_i < y_j$. They are said to be discordant, if $x_i > x_j$ and $y_i < y_j$ or if $x_i < x_j$ and $y_i > y_j$. If $x_i = x_j$ or $y_i = y_j$, the pair is neither concordant, nor discordant.

The Spearman's rho (ρ_S) coefficient is defined as

$$\rho_S = 1 - \frac{6\sum_{i=1}^{n} (x_i - y_i)^2}{n(n^2 - 1)} \tag{3.3}$$

Implicitly, the Pearson's correlation coefficient is a summary measure of linear relationship and it is based on normality. Hence, despite its simplicity and interpretability, it should be employed with caution. Indeed, a graphical check, such as a scatter plot, should go together with a correlation value to graphically check linearity of the relationship as well as the normality (approximately) of each of the two variables or at least low skewness. Alternatively, other measures can be considered such as the Spearman coefficient if the relationship is monotonic and/or if marginal distributions are not symmetric (Maindonald & Braun, 2010).

Given the fact that a dependence measure is a summary of the sample, the same value could be associated with many different patterns, and inversely, similar strengths of relationships can produce different coefficients. Even though it is common to consider Spearman and Kendall coefficients as alternatives to Pearson's coefficient, these coefficients measure a different type of relationship. Furthermore, the values of these coefficients should be tested especially for values close to zero.

In accordance with HFA aims, it is appropriate to consider, but not limited to, dependence measures that are related to copulas and not related to the margins. These measures are useful at this stage to have an idea about the dependence and guide or restrict the range of copulas to consider in the modeling step later.

Example

We present the different correlations for the Nottawasaga River as well as the results of the independence tests based on the Pearson's rho, the Kendall's tau, and the Spearman's rho. For comparison purposes, we considered the three couples from Q, V, and D (Table 3.1).

At a nominal value of 5%, we observe that (Q,V) are positively significantly dependent using the three measures, whereas (Q,D) are negatively dependent but with slightly lower values (significant at 10% for Pearson). However, (D,V) presented almost not significantly dependent. In particular for (Q,V), these values are in accordance with similar studies and also quantify the comments found for the contours and bagplot in the previous section.

TABLE 3.1 Pearson's rho, Kendall's tau, and Spearman's rho coefficients and the corresponding p-values of the test of independence for the Nottawasaga River bivariate series.

	Pearson's rho		**Kendall's tau**		**Spearman's rho**	
	Coefficient	***p*-value**	**Coefficient**	***p*-value**	**Coefficient**	***p*-value**
(Q,V)	0.38	0.0166	0.32	0.0037	0.44	0.0055
(Q,D)	−0.31	0.0617	−0.25	0.0259	−0.37	0.0214
(D,V)	0.29	0.0764	0.16	0.1498	0.21	0.1944

3.4 Outliers

Outliers are unusual values far from, or inconsistent with, the main body of data points. They could be observed values or not resulting from human errors among others. As in many other domains, outliers could be present in hydrological data series. The presence of outliers can affect the statistical exploratory analysis as well as the inference, such as on the estimated parameters of a given distribution and hence for HFA, the quantile estimation and the final decisions. Dealing with those outliers should be performed in the preliminary step of any statistical analysis.

Identifying outliers is an important statistical step to analyze datasets. In hydrologic data, outlier detection is a common problem which is considered in the univariate framework in several studies where it is generally treated very briefly (see details below). Once identified, the circumstances around these values have to be checked and investigated. Then, a decision should be taken to remove these values or accommodate them. For the latter, it is appropriate to consider robust approaches for further analysis. Removing data is the easiest solution but not always the most appropriate. There should be strong reasons for this action.

To detect outliers, a number of tests are available in the literature for univariate data, in particular, Anscombe, Dixon-Thompson, Rosner, and Grubbs. The latter, also called Grubbs and Beck test, is the most used and considered for hydrological applications as indicated, for example, in Rao and Hamed (2000). Note that these tests require normality of the data. More details about the univariate case can be found, for instance, in Barnett and Lewis (1998) and in hydrology in Rao and Hamed (2000), Panu and Ng (2017).

In extending univariate outlier detection methods to higher dimensions, various issues arise, such as limited visualization methods, inadequacy of marginal methods, lack of a natural order, and limited parametric modeling. To address and overcome such limitations, Dang and Serfling (2010) introduced nonparametric multivariate outlier identifiers based on depth functions, which can generate contours following the shape of the dataset.

In order to detect multivariate outliers, an outlyingness function and a threshold are needed. The former takes values usually in the interval [0, 1]. An outlyingness value of a point close to zero indicates that this point is close to the center, whereas a value near 1 indicates high outlyingness (far away). Furthermore, a threshold on outlyingness values is required to determine whether an observation is an outlier or not, that is, the minimum outlyingness value from which a datum is considered as an outlier. This approach is considered and applied to hydrological data, for instance by Chebana and Ouarda (2011) and Karahacane et al. (2020).

3.4.1 Outlyingness

Outlyingness treated here are based in the notion of depth function. The latter is briefly presented in Appendix. A depth outlyingness is a transformation of a depth function for a given distribution F and $x \in R^d$. A number of such outlyingness function are available including mainly:

Half-space:

$$O_{HD}(x,F) = 1 - 2HD(x,F) \tag{3.4}$$

Mahalanobis:

$$O_{MD}(x,F) = d^2_{A(F)}(x,\mu(F)) / \left[1 + d^2_{A(F)}(x,\mu(F))\right] \tag{3.5}$$

Projection:

$$O_{PD}(x,F) = PD(x,F)/[1 + PD(x,F)] \tag{3.6}$$

Spatial:

$$O_S(x,F) = \left\| E\left(\text{Sign}(x - X)\right) \right\| \tag{3.7}$$

Spatial Mahalanobis:

$$O_{SM}(x,F) = \left\| E\left[\text{Sign}\left(\mathbf{C}^{-1/2}(x - X)\right)\right] \right\| \tag{3.8}$$

where $HD(.,F)$, $d^2_{A(F)}(.,\mu(F))$, and $PD(.,F)$ are the associated depth functions given in Appendix, $\mu(F)$ is a location measure, $A(F)$ is a nonsingular matrix scale measure, $\|.\|$ is the Euclidean norm, X is F-distributed, and Sign(.) is the multidimensional sign function given by

$$\text{Sign}(x) = x/\|x\| \text{ if } x \neq 0 \quad \text{and} \quad \text{Sign}(0) = 0 \tag{3.9}$$

and $\mathbf{C}$ is any affine invariant symmetric positive definite $d \times d$ matrix. The matrix $\mathbf{C}$ could be the classical covariance matrix or the matrix obtained as the minimum covariance determinant (see Dang & Serfling, 2010).

3.4.2 Threshold

Selection of the appropriate threshold is related to false positive rate, denoted α_n. It represents the proportion of nonoutliers misidentified as outliers. A threshold is defined as the $(1-\alpha_n)$-quantile of the outlyingness values:

$$\lambda_n = F^{(-1)}_{O(X,F)}(1-\alpha_n) = F^{(-1)}_{O(X,F)}(1-\delta\varepsilon_n) = F^{(-1)}_{O(X,F)}\left(1-\frac{\beta\delta}{\sqrt{n}}\right) \tag{3.10}$$

where ε_n is the *true positive rate*, which represents the real theoretical proportion of outliers, $\delta = \alpha_n/\varepsilon_n$ is the ratio of false outliers and $\beta = \varepsilon_n\sqrt{n}$. Ideally, α_n has to be small compared to ε_n.

For instance, with $\delta = 0.05$, the ratio of false outliers is about 5% among the allowed ones. Assume that we allowed for $n\varepsilon_n = 20$ true outliers, the constant β becomes $\beta = n\varepsilon_n/\sqrt{n} = 20/\sqrt{100} = 2$ for $n = 100$. Hence, $\lambda_n = F^{-1}_{O(X,F)}(1 - 2 * 0.05/\sqrt{100}) = F^{-1}_{O(X,F)}(0.99)$ corresponds to the 0.99 quantile of the outlyingness values. For the multivariate normal distribution, formulas of these thresholds are given explicitly. Otherwise, these thresholds are not available in general where they can be approximated by those of the normal case.

Example

For the same station as in previous sections (at Nottawasaga River), Table 3.2 presents the bivariate outlier results for the couple (Q,V) as well as the univariate outlier testing by Rosner test. This test requires the data series to be normally distributed, which is usually not the case for hydrological data. Therefore they are transformed by the Box–Cox

TABLE 3.2 Univariate and bivariate detected outliers of Q and V for the Nottawasaga River.

	Years	Stats. value	Threshold	Conclusion
Univariate Peak Q	1990	2.82	3.03	Not outlier
	1999	2.56	3.01	Not outlier
	2002	2.13	3.00	Not outlier
Univariate Volume V	1999	2.12	3.03	Not outlier
	1985	2.05	3.01	Not outlier
	2009	2.03	3.00	Not outlier
Bivariate (Q,V)	1983; 1984; 1985; 1990; 1999; 2009; 2010	0.949	0.949 (Tukey)	Outliers
	1990	0.930	0.861 (Mahalanobis)	Outlier
	1990	0.941	0.936 (Spatial)	Outlier

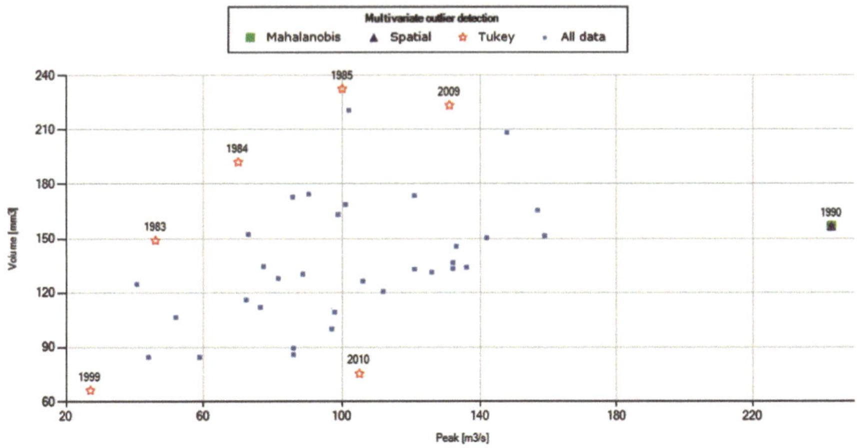

FIGURE 3.4 Bivariate detected outliers of Q and V with all data for the Nottawasaga River.

transformation with coefficients 0.40 and 0.47, respectively, for V and Q. Then, the Rosner test identified no outliers for each variable as shown in Table 3.2. Regarding the multivariate detection, by putting $\delta = 0.05$, the ratio of false outliers is about 5% among the allowed ones. Assume that we allowed for $n\varepsilon_n = 20$ true outliers (we obtain almost the same results with lower values), the constant β takes the value $n\varepsilon_n/\sqrt{n} = 3.20$ since here $n = 39$. Hence, the threshold λ_n corresponds to the 0.97 quantile of the outlyingness values. In this example, we considered Tukey, Mahalanobis, and spatial outlyingness measures.

Fig. 3.4 illustrates the multivariate outlying values in which the bivariate outliers can be detected for the (Q,V) couple.

The results show that in the univariate framework, no value is considered as outlier although the flood of 1990 is close to the critical value of Q. The years detected by the Tukey outlyingness are found on the threshold. However, the flood of 1990 is the only one detected by the three considered outlyingness measures. It was already identified in the bagplot as potential outlier. In a real-word study, an investigation should be done to find out the circumstances around this event. Since here it is an illustrative example, in the following we present results for both situations where this outlier is kept or removed.

3.5 Location measures

A location parameter either indicates where most of the data are located or it allows summarizing the data. This notion is useful in hydrology since it appears as a parameter in almost all commonly employed probability distributions. The usual location measures are the mean, the median, and the mode. The latter, as the most frequently observed value, is less commonly employed in hydrology. The mode can be used in HFA to detect heterogeneity in the sample. The reason behind its nonconsideration in hydrology could be the prior checking of the heterogeneity by appropriate tests or the prior separation of the events so that they are homogeneous. Note that it is recommended to consider more than one location measure for a given sample to avoid the possibility of missing important indication. For instance, by looking only at the mean, one might miss a possible heterogeneity which would be captured by the mode whereas the median is not sensitive to very high or low values in the sample. These issues are general and valid for the univariate as well as the multivariate settings (see Chebana & Ouarda, 2011 for more details and specific references).

3.5.1 Sample mean

The simplest and common location parameter is the arithmetic mean:

$$\mu_n = \frac{1}{n}\sum_{i=1}^{n} X_i \tag{3.11}$$

where $\mu_n = (\mu_{1,n}, \ldots, \mu_{d,n})$ is d-dimensional vector corresponding simply to the component-wise arithmetic means. Even though their advantages, the mean is sensitive to outliers. In the multivariate framework, unlike the mean, a number of ways are proposed to define the median.

3.5.2 Component-wise median

The component-wise median CM_n is a direct extension of the univariate median and it is given by

$$\mathrm{CM}_n = \left(\mathrm{med}\left(X_{1,1}, X_{2,1}, \ldots, X_{n,1}\right); \ldots; \mathrm{med}\left(X_{1,d}, X_{2,d}, \ldots, X_{n,d}\right)\right)' \tag{3.12}$$

where *med* is the univariate median. Similar to the univariate setting, CM_n is not affine equivariant and usually it is not an observation from the sample.

3.5.3 Depth-based median

This multivariate median is defined as the deepest point, that is the point for which the depth function is maximized. However, it is possible that more than one point maximize the depth function. Hence, in such situation, let $E \subseteq R^d$ be the set of points that maximize the considered depth function. The depth median is then the centroid of the polygon composed by the points in E.

Since each depth function should lead to its own depth median, the labels of these medians are given accordingly such as Tukey median, Oja median, or Liu median. Even though any other depth function could also be used, the above are the most studied in the literature. Indeed, medians based on Tukey and Oja depths have suitable properties whereas the one based on Liu depth is not studied yet. For instance, the set E is convex with Tukey depth whereas the set E is a single point for the Oja median, if n is even.

3.5.4 Spatial median

It is defined, using the Euclidean norm $\|.\|$, as

$$\text{SpMed} = \frac{1}{n} \underset{x \in R^d}{\operatorname{argmin}} \sum_{i=1}^{n} \|x - X_i\| \tag{3.13}$$

where $\operatorname{argmin}_{t \in A} \varphi(t)$ is the minimizer of the function $\varphi(.)$ over a set A.

3.5.5 α depth-trimmed mean

The classical sample mean uses all available observations. However, its performance is severely affected by the presence of extreme values and outliers, especially for small size samples, which is usually the case in HFA. On the other hand, the median is known not to be affected by extreme values or outliers. To estimate a population location, the trimmed sample mean has been widely accepted as a more robust estimator than the untrimmed mean, in the sense that its behavior is not affected by the outliers and heavy tails of population distributions.

Given a coefficient $0 \le \alpha \le 1$ and a depth function, the α depth-trimmed mean corresponds to the sample mean based on the $100(1-\alpha)\%$ deepest points. Formally, based on the sorted sample $X_{[1]}, \ldots, X_{[n]}$, we define the R^d-valued function ξ_n on $[0, 1]$ as

$$\xi_n(t) = X_{[i]} \text{ if } \frac{i-1}{n} < t \le \frac{i}{n} \text{ and } \xi_n(0) = X_{[1]} \tag{3.14}$$

where $\overline{\xi}_n(t)$ is the average over the *de*-class values in which $\xi_n(t)$ is contained. For $0 < \alpha < 1$, if $n\alpha$ is an integer, then the α trimmed mean is $\text{TM}_n = \frac{1}{n(1-\alpha)} \sum_{i=1}^{n(1-\alpha)} X_{[i]}$.

It is common to consider values of α such as 0.25. However, as particular cases, when $\alpha = 1$, then TM_n becomes the median whereas for

$\alpha = 0$, it coincides with the usual mean. The trimmed mean has a number of interesting properties. For instance, it is a resistant estimator of location, as it is not strongly influenced by outliers. The trimmed means are particular examples of weighted means with particular choice of the weights. Much research on the performance of these location estimators has been conducted on distributions with heavy tails. This is appropriate and of particular interest for hydrological applications, especially regarding risk assessment. Despite its advantages, the trimmed mean is not commonly used in hydrology.

Given the variety of the available location measures and estimators, their performances are compared in the univariate as well as in the multivariate settings based on different criteria. In the univariate framework, several studies highlighted the superiority of the trimmed mean. However, in the bivariate case, a numerical comparison study provided the following results. Overall, medians were shown to be more robust location parameters than means. More precisely, in terms of robustness and accuracy, the spatial median stands as the best location parameter, followed by Oja and Tukey medians, which is considered as a first group. In a second group, in terms of robustness, we find the Liu and the component-wise medians. Trimmed means (for $\alpha = 0.05$, 0.10 with Tukey and Liu depths) are in a third group, and in the last group we have the sample mean. All the above location measures, except for the Liu and Oja medians, can be obtained in higher dimensions, though sometimes under approximations.

Example

For the Nottawasaga River, Fig. 3.5 present the different location measures for the (Q,V) series. We focus only on Tukey depth function among the three available ones, for presentation clarity and since the results are very similar. The trimmed measure is calculated with $\alpha = 0.25$. In addition, for comparison purposes, we considered including all data versus excluding the detected outlier 1990. We observe in general that all the location measures are located in the center of the data cloud. However, the Tukey median is located more in the down-left. When excluding the outlier, the mean-based measures are the most affected (especially in terms of the peak variable) whereas the median-based ones are almost not affected.

3.6 Scale measures

After summarizing the data through location measures, the next step is to evaluate the dispersion of the sample around a location measure. Scale parameters are useful to this end. Matrix-valued and

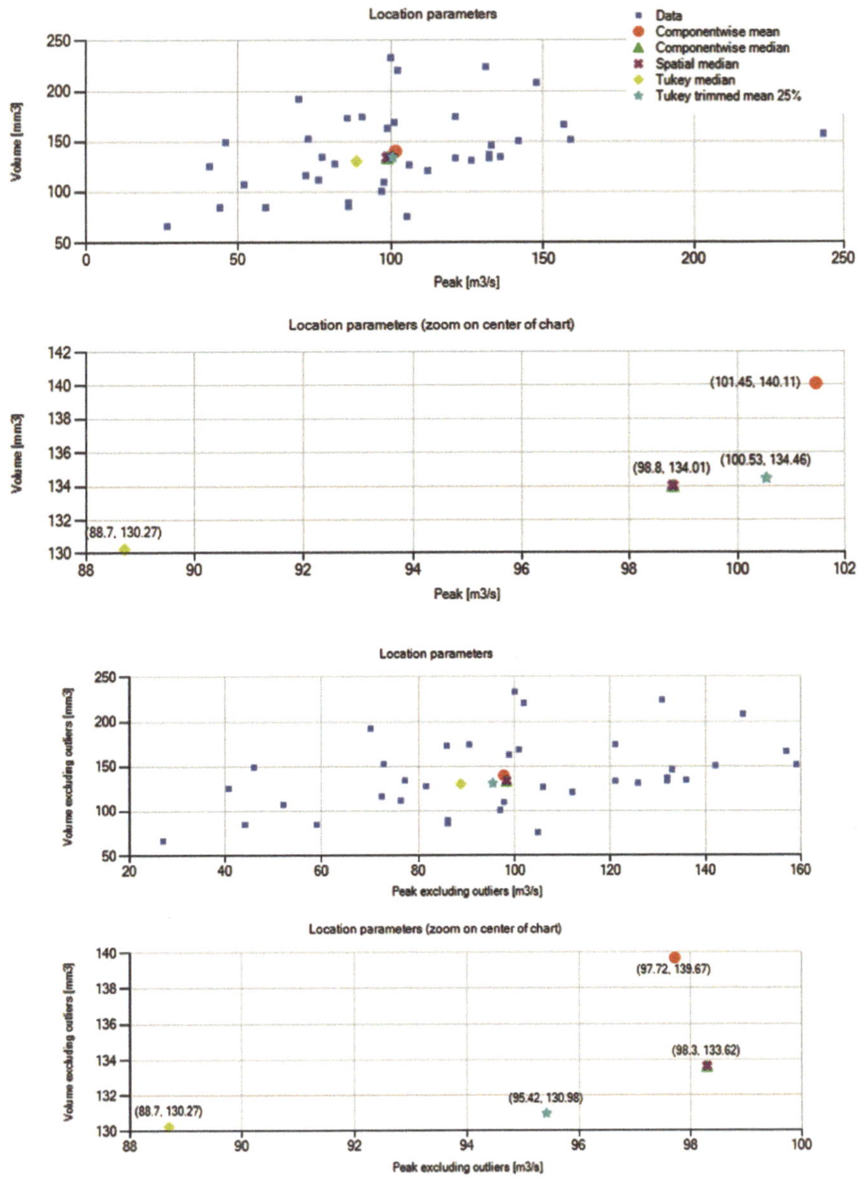

FIGURE 3.5 Location measures with Tukey depth function including (left) and excluding (right) the detected outlier 1990 for (Q,V) at Nottawasaga River.

scalar-valued represent two types of multivariate scale parameters. First, we focus on the α-trimmed scale measure since it is general and includes other measures as special cases similarly as in the location measures.

3.6.1 α-trimmed sample dispersion matrix

Given a center-outward ranking of data resulting from a given depth function, the corresponding definition is similar to that of the trimmed mean, where we replace $\xi_n(t)$ by the function $S_n(t)$ defined by Liu et al. (1999):

$$\mathbf{S}_n(t) = \left(X_{[i]} - \nu_n\right)\left(X_{[i]} - \nu_n\right)' \quad \text{if } \frac{i-1}{n} < t \leq \frac{i}{n}, \quad \text{and } \mathbf{S}_n(0) = \mathbf{0}_{d\times d} \tag{3.15}$$

where v_n is the sample's deepest point and $\mathbf{0}_{d\times d}$ is the $d \times d$ matrix with null elements. Given $0 \leq \alpha < 1$, if $n\alpha$ is an integer, the *α-trimmed-dispersion matrix* is given by

$$TD_n = \frac{1}{n(1-\alpha)} \sum_{i=1}^{n(1-\alpha)} \overline{\mathbf{S}}_n\left(\frac{i}{n}\right) \tag{3.16}$$

where $\overline{\mathbf{S}}_\mathbf{n}$ indicates the average of $\mathbf{S}_\mathbf{n}$ over all *de*-classes to which $X_{[i]}$ belongs. For $\alpha = 1$, we define TD_n as the zeros matrix and for $\alpha = 0$, it coincides with the usual covariance matrix.

Note that the matrix-valued scale allows an easy comparison of dispersion between dimensions and can reveal more information. However, it could not be the appropriate tool to measure the overall dispersion of the distribution. In this sense, although it can be seen as reduction of the information, taking a norm of the scale matrix (3.16) is an alternative.

3.6.2 Scalar form of scale

Scalar-valued scale can be seen as an information reduction. Hence, to overcome this limitation, it is proposed to plot these values as a curve. To this end, given a depth function, let $\mathrm{Sc}_n(p)$, $0 \leq p \leq 1$ be the function that returns the volume of the central region $C_{n,p}$ composed of the $\lceil np \rceil$ deepest points, where $\lceil a \rceil$ is the smallest integer, larger or equal to a.

The graphic of $(\mathrm{Sc}_n(p),\ p)$ is an evaluation of the expansion of the central region $C_{n,p}$ with respect to p. The obtained curve $\mathrm{Sc}_n(.)$ is interpreted as follows: "if the scale curve of a distribution G is consistently above the scale curve of another distribution F, then G has a larger scale than F" (Liu et al., 1999).

Example

The results of the scale measures for the (Q,V) series of the Nottawasaga River for $\alpha = 0.25$ are summarized in Table 3.3 with different depth functions. In addition, Fig. 3.6 shows the scalar scale plot with Tukey depth function as well including/excluding the detected outlier (as for location).

TABLE 3.3 Dispersion matrix measures for peak and volume at Nottawasaga River.

		All data		Excluding outlier 1990	
		Peak Q (m^3/s)	Volume V (Mm^3)	Peak Q (m^3/s)	Volume V (Mm^3)
Dispersion		1628.18	626.85	1116.40	578.45
		626.85	1660.47	578.45	1697.67
Trimmed dispersion 25%	Tukey	1081.00	555.51	814.16	368.49
		555.51	945.58	368.49	1125.81
	Oja	1132.43	433.02	1147.46	459.03
		433.02	735.04	459.03	760.96
	Liu	1039.67	463.96	851.65	351.99
		463.96	1030.58	351.99	982.58
	Mahalanobis	1122.27	327.04	1050.37	432.15
		327.04	678.94	432.15	760.26

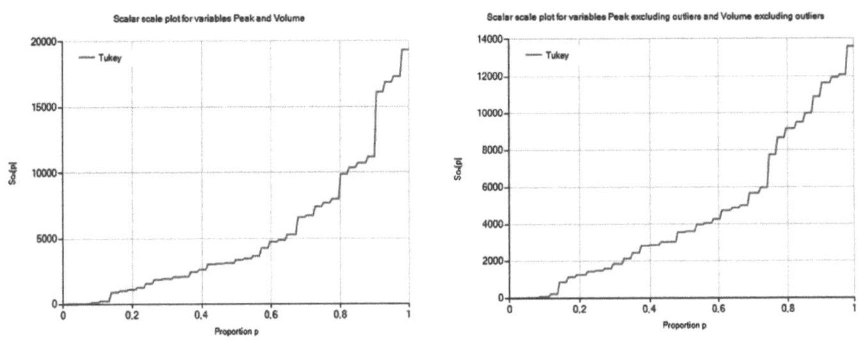

FIGURE 3.6 Scalar scale plot for peak and volume at Nottawasaga River including (left) or excluding (right) the detected outlier 1990.

From Table 3.3, we observe that the dispersion is more affected regarding the peak Q when excluding the outlier, which is not the case for V. In addition, comparing dispersion with the 25% trimmed dispersion, the latter is less affected. From Fig. 3.6, we can observe that by excluding the outlier, the data become less spread. This could be seen as a simpler way to compare than with the previous matrices.

3.7 Asymmetry

Generally in hydrology, and in particular in HFA, distributions (univariate) are not symmetric and are one-side heavily tailed (Helsel & Hirsch, 2002). This is because hydrologic data are generally skewed where extreme values often occur in one direction of the tail of the distribution. In situations similar to the above, when symmetry fails to hold, it is of interest to characterize the skewness. The latter can be seen as a measure of the nature or direction of the departure from symmetry (Serfling, 2006). In the multivariate setting, a useful introduction regarding these notions can be found in Fang et al. (1990).

As for other notion in the multivariate setting, more than one definition of multivariate symmetry may be expressed. Generally speaking, symmetry can be expressed in terms of invariance of the distribution of a "centered" random vector $X - \theta$ in R^d under a suitable family of transformations. Typical examples of types of symmetry include spherical, elliptical, antipodal, and angular. All of them reduce to the classical univariate symmetry. In the following, before providing an illustrative example, the definition of each symmetry is presented as well as how it can be empirically evaluated (Liu et al., 1999; Serfling, 2006). For the latter, depth-based tools are presented. Recall that depth function is briefly described in Appendix.

3.7.1 Spherical symmetry

It is defined as "The distribution of the random variable X is said to be spherically symmetric about the point c if the distributions of $(X - c)$ and $\mathbf{U}(X - c)$ are identical, for any orthonormal matrix $\mathbf{U}$." A matrix $\mathbf{U}$ is called orthonormal if and only if $\mathbf{U}\mathbf{U}' = \mathbf{U}'\mathbf{U} = I$ where $\mathbf{U}'$ is the transpose of $\mathbf{U}$ and I is the identity matrix. From its definition, this symmetry represents a rotation of X about the point c. In terms of the probability density function of X, when it exists, it is of the form $g((x - c)'(x - c))$ for a nonnegative real-valued function g. Typical examples of such distributions include multivariate normal, student, and logistic distributions.

To empirically evaluate the spherical symmetry for a given sample of size n, we consider, for a given depth function, the smallest enclosing d-sphere that contains the $\lceil np \rceil$ deepest points for $p \in [0, 1]$. The proportion of sample points falling in this sphere is denoted by Sph(p) as a function of p. This function is increasing and ranges between p and 1. As an indicator of spherical skewness, the area Δ_n between the curve $y = \text{Sph}(p)$ and the diagonal line $y = p$ plays this role. In particular, when Δ_n is close to zero, it means the curve Sph(.) is close to the diagonal (i.e., Sph $(p) = p$) and hence the corresponding sample can be considered as a perfectly spherical symmetric sample.

3.7.2 Elliptical symmetry

It is defined as "The distribution of the random variable X is said to be elliptically symmetric about a certain point c if there exists a nonsingular matrix **V** such that **V**X is spherically symmetric about c." A general form of the corresponding probability density function of X is $|V|^{-1/2}g\left((x-c)'V^{-1}(x-c)\right)$. One of the particular examples in this class is the multivariate normal distribution with $\Sigma = V'V$ as a covariance matrix. As can be expected, the corresponding contours of this probability density function are of elliptical shape.

One can observe that from their respective definitions, the elliptical symmetry is similar to the spherical one after a standardization with the matrix **V**. Hence, in terms of its empirical evaluation, for an elliptical skewness, data are first standardized using the scale matrix of the $\lceil np \rceil$ deepest points for $p \in [0,1]$. Then, based on the transformed data, we proceed as in spherical symmetry by evaluating and plotting the function Sph(p). In the case of elliptical skewness, the interpretation of the curves associated is similar to that of the spherical skewness.

3.7.3 Antipodal symmetry

It is defined as "The distribution of the random variable X is said to be antipodally symmetric about the point c (if such a point exists) if the distributions of $(X-c)$ and $-(X-c)$ are identical." This symmetry can be seen as the most direct extension of the usual univariate symmetry. It is also called reflective or diagonal. The corresponding probability density function f is such that $f(x-c) = f(c-x)$.

To evaluate this symmetry, we consider a depth function and a location parameter μ. Let $C_{n,p}$ be the pth central region. We denote $Ca(p)$ the proportion of the $\lceil np \rceil$ deepest points falling in the intersection of $C_{n,p}$ and its reflection (the reflection is about μ for p in (0, 1)). From the definition, we have $0 \le Ca(p) \le \lceil np \rceil / n$. For a perfect antipodal symmetric sample, it would suggest that $Ca(p)$ reaches almost its upper limit, that is, $Ca(p) = \lceil np \rceil / n \approx p$. As for previous evaluations, the area between the diagonal line $y = x$ and the curve $y = Ca(x)$, for $x \in [0, 1]$, is used as indicator to measure antipodal skewness where a large area corresponds to a larger deviation from antipodal symmetry.

3.7.4 Angular symmetry

It is defined as "The distribution of the random variable X is said to be angularly symmetric about the point c if, conditional on $X \neq c$, the distributions of $(X-c)/\|(X-c)\|$ and $-(X-c)/\|(X-c)\|$ are identical." One of the features of angular symmetry about c is that any hyperplane passing through c divides the whole space R^d into two half-spaces with equal probabilities 0.5 (if the distribution is continuous).

Given a multivariate sample for which we want to measure angular symmetry, first we identify the deepest point ν_n according to a given depth function. Then, we evaluate the Tukey depth of the deepest point ν_n with respect to the restricted data in the pth central region $C_{n,p}$ for each $p \in [0, 1]$. The degree of the antipodal symmetry is measured by the deviation of the obtained curve, denoted $h(p)$, from the x axe. Note that the value of Tukey depth of the deepest point should be 0.5 under angular symmetry. According to Liu et al. (1999), the obtained values and curves can be interpreted as: "[...] the deviation of the half-space depth at the deepest point from the value 0.5 is a measure of the departure from angular symmetry of the empirical distribution determined by the sample points within each level set." For convergence reasons, it is suggested to consider only the part of the curve $h(.)$ with p larger than 0.4 where the curve stabilizes.

It is worth to mention that the above symmetry notions are related to each other. Indeed, they can be ranked from more to less restrictive from spherical, elliptical, and antipodal to angular symmetry. In this sense, for instance, if a variable X is elliptical symmetric, then it is antipodal as well as angular. Once defined and evaluated, it would be of interest to conduct hypothesis testing regarding symmetry. This is a recent and current topic of research in the multivariate framework not covered in this book.

Example

Fig. 3.7 presents the different symmetry plots of the (Q,V) series for the Nattawasaga River in both cases when the detected outlier 1990 is excluded or included. We used only the Tukey depth not only for clarity of the figures but also since the results with the different depth functions are very similar in general. We can observe that by excluding the outlier, the data cloud become more spherical and more elliptical but almost not affected regarding antipodal and angular symmetries.

3.8 Kurtosis

Kurtosis is an important concept in hydrology and HFA, especially when the focus is often on the tail of the distribution where extreme events occur. On the other hand, characterizing kurtosis is the natural step after treating location, scale, and asymmetry features. However, kurtosis is more complex to characterize and interpret since it is related to notions like spread, peakedness, and tailweight, and even with asymmetry (Wang & Serfling, 2005).

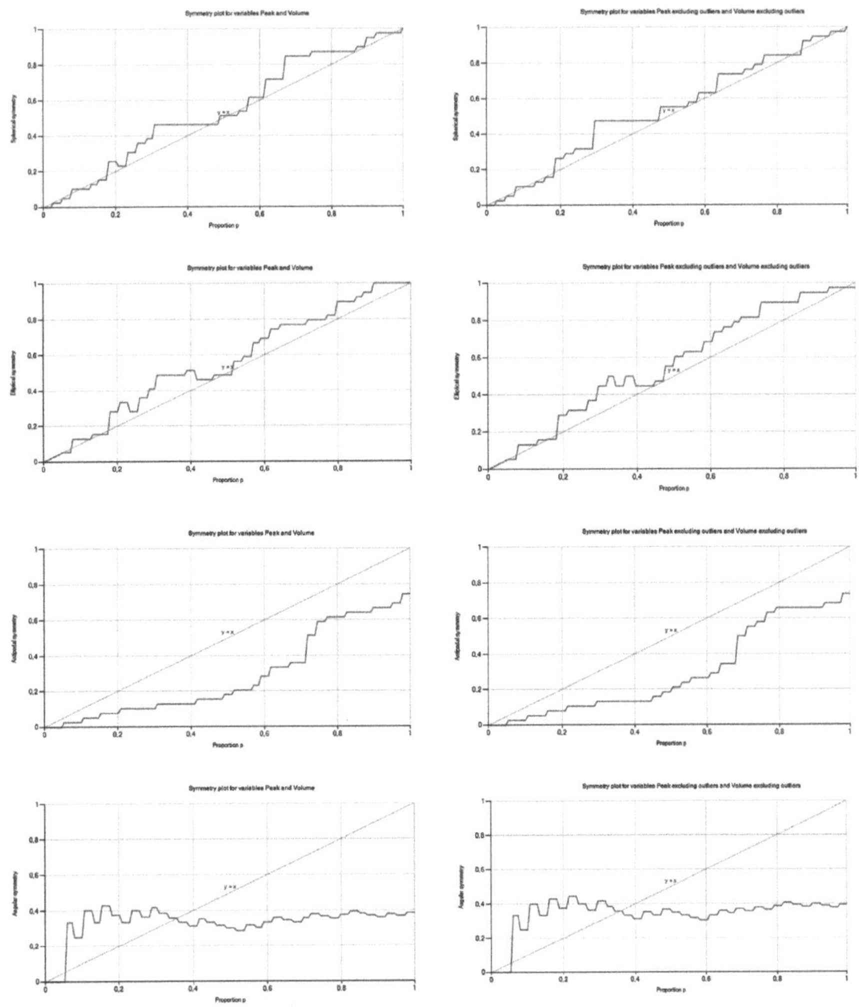

FIGURE 3.7 Symmetry plot with different symmetries for peak and volume at Nottawasaga River including (left) or excluding (right) the detected outlier 1990.

One of the definitions of kurtosis is based on a ratio of two scale measures, that is, scale of the whole data and scale of the central part. In the following, we present four kurtosis measures. The reader is referred to Liu et al. (1999) and Wang and Serfling (2005) for more details.

3.8.1 Lorenz curve of Mahalanobis distance

Let $\mathbf{S}_n$ be a nonsingular scale matrix, and a given depth function, we have the following versions of the Lorenz curve:

$$L(p) = \frac{\sum_{i=1}^{\lceil np \rceil} Z_i}{\sum_{i=1}^{n} Z_i} \text{ and } L^*(p) = \frac{\sum_{i=1}^{\lceil np \rceil} Z_i / \lceil np \rceil}{\sum_{i=1}^{n} Z_i / n} \text{ for } 0 < p \leq 1 \tag{3.17}$$

where

$$Z_i = \left(X_{[i]} - \nu_n\right)' \mathbf{S}_n{}^{-1} \left(X_{[i]} - \nu_n\right), \text{ for } i = 1, 2, \ldots, n \tag{3.18}$$

is the Mahalanobis distance and ν_n is the deepest point. As an example, $\mathbf{S}_n$ can be the matrix given in (3.16) or the covariance matrix.

For the limiting cases of $p = 0$ and 1, we define $L(0) = L^*(0) = 0$ and we have $L(1) = L^*(1) = 1$. As in skewness measures, the area corresponding to the surface between the curve $y = L(x)$ or $y = L^*(x)$ and the diagonal line $y = x$ is evaluated. The interpretation is similar for both cases: a large area corresponds to a high degree of peakedness and tailweight, and inversely a small area corresponds to heavy shoulders. Even though the curves L^* and L have the same interpretation, from their definitions, the area computed from L^* should be more noticeable than the one from L. Hence, it is more effective to compare samples using L^* than L.

3.8.2 Shrinkage plot

It is based on the shrinkage of the boundary of the pth central region $C_{n,p}$ towards its center by a fixed coefficient s, $0 < s < 1$ resulting in region $C^s_{n,p}$. For fixed s, the function $a_s(p)$ of the fraction of observations in $C^s_{n,p}$ is to be plotted. The value $s = 0.5$ was proposed by Liu et al. (1999). For a fixed s, heavier tails correspond to higher values of $a_s(p)$ especially for large p.

3.8.3 Fan plot

It is a collection of an arbitrary number of curves aiming to evaluate kurtosis. Each curve is associated with a given value of $p \in [0, 1]$ where the subsample Sam(p) is formed by the $\lceil np \rceil$ deepest points (in the central region $C_{n,p}$). For $t \in [0, 1]$, let $C_n(p, t)$ be the area of the tth convex hull of Sam(p) composed by $100t$ % of the deepest observations. Then, the function $b_p(t)$ for $t \in [0, 1]$ is given by

$$b_p(t) = \frac{\text{volume}\left[C_n(p, t)\right]}{\text{volume}\left[C_n(p, 1)\right]} \text{ if } C_n(p, 1) \neq 0 \text{ and } b_p(t) = 0 \text{ otherwise} \tag{3.19}$$

Intuitively, a fan plot may be regarded as a comparison of areas between central (corresponding to low values of p), shoulder (corresponding to middle values of p), and tail regions (corresponding to high values of p). A more spread out fan plot indicates that the corresponding distribution is heavy tailed since $b_p(t)$ becomes smaller. This way to measure kurtosis requires a large amount of data since the data size is reduced in two stages (with p and then with t). Hence, this measure could be not appropriate for current HFA applications.

3.8.4 Quantile-based measure

It is based on the function $k_C(.)$, which is expressed as

$$k_C(r) = \frac{V_C\left(\frac{1-r}{2}\right) + V_C\left(\frac{1+r}{2}\right) - 2V_C\left(\frac{1}{2}\right)}{V_C\left(\frac{1+r}{2}\right) - V_C\left(\frac{1-r}{2}\right)} \text{ for } 0 < r \leq 1 \text{ and } k_C(0) = 0 \tag{20}$$

where the function $V_C(r)$ is the volume of a central set $C(r)$. The set $C(r)$ is defined as the inner set, with probability r, delimited by contours of a given depth function. For the latter, Tukey depth function has been used and any affine invariant depth function can also be used. The measure $k_C(r)$ can be seen as the difference of the volumes of two regions A and B divided by their total volume where $A = C(1/2) - C(1/2 - r/2)$ and $B = C(1/2 + r/2) - C(1/2)$. Note that the boundary associated to the region $C(1/2)$ represents the "shoulders" of the distribution and it separates the "central part" from the corresponding "tail part."

To interpret the obtained curves, unlike the other kurtosis measures discussed above, the quantile-based measure needs some prior knowledge about the distribution of the sample. Indeed, Wang and Serfling (2005) indicated that if the attention is confined to a class of distributions for which either F is unimodal, F is uniform, or $1 - F$ is unimodal, then, for any fixed r, a value of F near $+1$ suggests a peakedness, a value near -1 suggests a bowl-shaped distribution, and a value near 0 suggests uniformity. Increasing values of k_C (.) indicate that the probability mass is greater in the center than in the tails.

Example

Fig. 3.8 presents a comparison between the Kurtosis plots of the (Q,V) couple for the Nottawasaga River when the detected outlier 1990 is excluded or included. We used only the Tukey depth for clarity of the figures but also because the results with the different depth functions are very similar in general (except for the quantile-based measure).

Based on the L-measure, there is almost no difference by excluding or including the outlier. However, using L*, the peakedness degree and

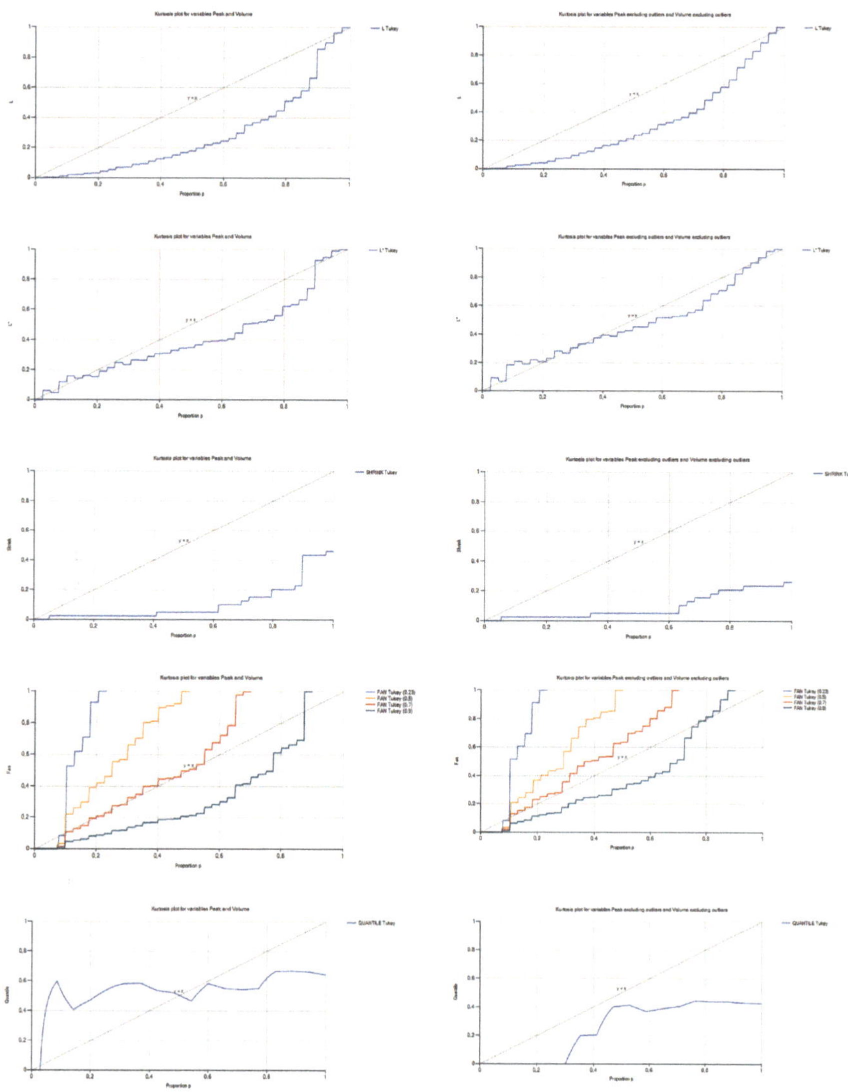

FIGURE 3.8 Kurtosis measures for peak and volume at Nottawasaga River including (left) or excluding (right) the detected outlier 1990.

tailweight decreased when excluding the outlier. This confirms, as mentioned above, that L* is more appropriate than L to measure kurtosis. Using shrink and fan measures, we observe similar effect as with L*. Regarding the quantile measure, at this stage we only say there is an effect of excluding the outlier but cannot interpret since we do not have any information about the distribution.

References

Anderson, T. W. (1984). *An introduction to multivariate statistical analysis*. New York: Wiley.

Barnett, V., & Lewis, T. (1998). *Outliers in statistical data*. Chichester: Wiley, [etc.].

Chebana, F. (2013). *Multivariate analysis of hydrological variables. Encyclopedia of environmetrics* (2nd ed.). John Wiley & Sons. Available from http://doi.org/10.1002/9780470057339.vnn044.

Chebana, F., & Ouarda, T. B. (2011). Depth-based multivariate descriptive statistics with hydrological applications. *Journal of Geophysical Research: Atmospheres, 116*(D10).

Dang, X., & Serfling, R. (2010). Nonparametric depth-based multivariate outlier identifiers, and masking robustness properties. *Journal of Statistical Planning and Inference, 140*, 198–213.

Fang, K.-T., Kotz, S., & Ng, K. W. (1990). *Symmetric multivariate and related distributions*. London: Chapman & Hall.

Helsel, D.R., & Hirsch, R.M., Geological Survey (U.S.) (2002). *Statistical methods in water resources*. Reston, VA: U.S. Geological Survey.

Karahacane, H., Meddi, M., Chebana, F., & Saaed, H. A. (2020). Complete multivariate flood frequency analysis, applied to northern Algeria. *Journal of Flood Risk Management, 13*(4)e12619.

Liu, R. Y., Parelius, J. M., & Singh, K. (1999). Multivariate analysis by data depth: Descriptive statistics, graphics and inference. *Annals of Statistics, 27*, 783–858.

Maindonald, J., & Braun, W. J. (2010). *Data analysis and graphics using R: An example-based approach*. Cambridge University Press.

McLeod, A. I., & Provost, S. B. (2006). *Multivariate data visualization. Encyclopedia of environmetrics*. John Wiley & Sons, Ltd.

Panu., & Ng. (2017). Outlier analysis and infilling of missing records in hydrological data Chapter 16. In V. P. Singh (Ed.), *Handbook of applied hydrology*. New York: McGraw-Hill Education.

Rao, A. R., & Hamed, K. H. (2000). *Flood frequency analysis*. Boca Raton, FL: CRC Press.

Serfling, R. (2006). Multivariate symmetry and asymmetry. In S. Kotz, N. Balakrishnan, C. B. Read, & B. Vidakovic (Eds.), *Encyclopedia of statistical sciences* (pp. 5338–5345). Wiley.

Wang, J., & Serfling, R. (2005). Nonparametric multivariate kurtosis and tailweight measures. *Journal of Nonparametric Statistics, 17*, 441–456.

Zhang, L., & Singh, V. (2006). Bivariate flood frequency analysis using the copula method. *Journal of Hydrologic Engineering, 11*(2), 150–164.

CHAPTER

4

Checking basic assumptions for multivariate hydrological frequency analysis

In this chapter, we treat the testing step within the framework of the hydrologic frequency analysis (HFA), which is briefly described in Chapter 2. In this chapter, the corresponding techniques and methods will be presented with more details and some examples. Nonstationarity, heterogeneity, and serial dependence need to be tested before the modeling step in a multivariate HFA. The testing step is important to ensure the respect of the basic assumptions. Hence, the selected models will be appropriate. These statistical tests are general and can be adapted and applied to a variety of hydrometeorelogical variables such as floods, storms, heat waves, and draughts, as well as to other fields.

4.1 Introduction and general considerations

A number of essential assumptions are required in order to accomplish most statistical analysis in several application fields, including hydrology and climatology. Indeed, particularly for HFA, stationarity, homogeneity, and independence are those assumptions to be tested. Actually, it is advisable to apply more than one test for each feature to check, in order to obtain reliable information about these features. As discussed in Chapter 2, in the multivariate setting, the aspect of testing the required hypotheses is generally ignored and the focus is made on modeling. This preliminary step attracted less attention in multivariate HFA (e.g., Gu et al., 2018).

The assumptions of stationarity and homogeneity, in the context of HFA, are closely related but not the same. Indeed, the statement of the null and alternative hypotheses to be tested is one of the differences.

DOI: https://doi.org/10.1016/B978-0-323-95908-7.00002-5

Homogeneity can be seen as the collected data do not represent a mix of different subpopulations where it is not required to respect the chronological order of the data. This is not the case for stationarity. A time series of observed data is commonly considered stationary and homogenous if it is free of *trends* (gradual monotonic changes), *shifts* (jumps), or *cycles* (periodicity) (Salas, 1993). The practical interpretation of these properties is that the data series comes from the same population and that the series' behavior does not change over time, assuming that the characterizing statistical parameters remain constant over time.

It is possible and more and more likely, for a number of reasons such as in a climate change context, that these assumptions would no longer hold. Therefore, conventional HFA could lead to invalid or suspicious results and conclusions. In order to be more realistic under such situations, it is important to consider suitable approaches which incorporate any nonsatisfied feature (nonstationarity, nonhomogeneity, and/or non-independence). As in a number of other fields, detecting changes in hydrological processes is essential for hydrological risk analysis and ecological system protection, among other water resources management activities. Nonstationarity, especially trends, was the object of a large number of hydrometeorological studies such as for floods and precipitations. Beside natural causes, man-made activities are also among the causes of these changes including, for instance, construction of dams, land use change, and urbanization. Indeed, changes in hydrological variables can be caused by changes in climate variables which in turn are potentially affected by the increase in the emission of greenhouse gases into the atmosphere. In addition, different forms of nonstationarities, such as jumps, trends, or cycles, can be associated with different causes. An abrupt change in basin or river system dynamics due, for instance, to the construction of a dam may cause jumps. Urbanization and changes in land use may lead to trend in streamflows. Commonly, long-term climatic fluctuations are related to cyclic nonstationarities.

The homogeneity may be not fulfilled in situations caused by natural or anthropogenic actions, such as deforestation, dam construction, and urbanization. In particular, floods can be the result of different hydrometeorological processes such as from storms, rainfall, and snowmelt. Another context where the homogeneity could fail is when the regime of a river downstream is formed from the confluence of two or more different subwatersheds. The fail of homogeneity, or heterogeneity, can be seen as a shift in the data. The presence of the latter is emphasized in several hydrometeorological studies such as for precipitations and floods.

Multivariate setting has two types of dependence: serial- and cross-dependence. The former deals with multivariate observations, whereas the dependence between the component variables is called cross-dependence. The (serial) dependence means, in simple terms, that any particular observation is not independent of its previous observations. The dependence in samples of hydrometeorological variables is usually characterized by the persistence phenomenon. In practice, for hydrometeorological data, the assumption of independence does not hold when the sampling frequency is quite high, or in general when the extreme values are obtained from the peaks-over-threshold approach (e.g., Khaliq et al., 2006).

The hypothesis testing step, in multivariate HFA, has a significant impact on the subsequent steps of the analysis such as the selection of the appropriate model. Indeed, if a trend as a nonstationarity in the data is detected, the appropriate model should include possible trends in some or all parts of the multivariate distribution (copula and margins). Hence, not performing the testing step or ignoring its results may lead to inaccurate model. As a consequence, wrong results could be obtained leading to inappropriate decisions. As an example, inappropriate model for the design of hydrological engineering planning and water structures could cause loss of human lives and property associated to underestimation, or high construction costs associated to overestimation. In conclusion, it is necessary to include hypothesis testing step in the multivariate HFA (see e.g., Chebana, 2013).

A wide variety of parametric methods, to deal with nonstationarity and heterogeneity, has been developed mainly for time series analysis with a range of application fields such as finance and econometrics. Linear regression is one of the most common parametric technique to investigate monotonic trends or shifts. To consider regression analysis, usually it requires normal (or prior known) distribution and independent data. However, hydrometeorological time series generally display particular features that make nonparametric tools as favorite ones for this step. First, in particular for HFA applications, the nonparametric tests are more appropriate since the testing step is prior to the modeling one. One of the drawbacks of the parametric trend tests is related to the number of involved parameters. This number becomes more important in the multivariate framework, which grows rapidly with the dimension as well as with the shape of the trend (e.g., linear, quadratic). Despite that, for normal data, parametric tests perform slightly better than their nonparametric counterparts, the latter are preferably applied in hydrology. In Oja and Randles (2004), the reader can find more details regarding parametric versus nonparametric tests. Hence, in the following, we focus on nonparametric tests.

In testing hypothesis, the p-value is important and is commonly used in practice as a simple criterion in order to accept or reject the null hypothesis. In principle, the evaluation of the p-value is based on the distribution of the test statistic. According to the complexity of the test statistic, if this distribution is unknown, asymptotic approximation or sampling methods can be considered. A brief and general summary regarding the p-value is given in Appendix.

Let $(X_i)_{i=1,\ldots,n} = \left(X_i^{(1)}, X_i^{(2)}, \ldots, X_i^{(d)}\right)'_{i=1,\ldots,n}$ be a sample of size n from a continuous stochastic process with dimension d ($d \geq 1$, $n \geq d$), where $'$ denotes the matrix transpose. As an example, we have for floods the variables annual flood peak Q, volume V, and duration D where here $d = 3$. The vector $x_i = \left(x_i^{(1)}, x_i^{(2)}, \ldots, x_i^{(d)}\right)'$ denotes the observation from X_i at time i in chronological order, that is, we start observing the process at time $i = 1$ and the last observation is obtained at time $i = n$. In the following, the term "time" is employed for simplicity and can refer to any other covariate.

4.2 Stationarity

In HFA, a data series is considered as statistical stationarity if it does not exhibit any significant changes over time; more precisely, when the probability distribution of the data is not significantly changing over time. It can also be for some of the distribution features such as location and scale. On the other hand, in statistical hydrology, several forms of nonstationarity are generally considered: the presence of trends, shifts, or a periodic behavior.

As a formal definition, a multivariate series $(X_i)_{i=1,2,\ldots,}$ is *strictly stationary* if X_i is identically distributed for all times $i = 1, 2, \ldots$, that is, the probability distribution does change over time (or covariate). The series $(X_i)_{i=1,2,\ldots,}$ is *weakly stationary* (or stationary in covariance), if it satisfies the following conditions:

1. $E(X_i) = \mu < \infty$ for $i = 1, 2, \ldots$ where μ is a vector of constant parameters,
2. $c_{i,i+k}^{(u,v)} = \text{cov}\left(X_i^{(u)}, X_{i+k}^{(v)}\right) = \text{cov}\left(X_j^{(u)}, X_{j+k}^{(v)}\right) < \infty$ for $k \in \mathrm{N}, i, j = 1, 2, \ldots,$ and $u, v = 1, \ldots, d$.

From their definitions, strict stationarity is clearly more restrictive than the weak one. Indeed, when the whole distribution remains the same over time, all corresponding moments, including the mean and the covariance, also remain the same. In a variety of applications, including hydrometeorology, the weak stationarity is usually employed. Since we focus on trend and its detection based on hypothesis testing, we consider the following definition.

A *trend* in a time series is defined as a monotonic change at least in one component of the vector series. Assume that the multivariate series $(X_i)_{i=1,\ldots,n}$ can be represented in an additive form as:

$$X_i = Y_i + T_i, \quad i = 1, \ldots, n \tag{4.1}$$

where Y_i is a stationary series, and the function $T(i) = T_i \in R^d$ called trend function is a deterministic function, which is monotonic in at least in one component $T^{(u)}, u = 1, \ldots, d$. As a first simple example of a trend is when $T(.)$ is a linear function. For trend analysis in HFA, we usually assume data to be serially independent which is reasonable in a number of situations such as for the case of maximum annual series.

Usually gradual monotonic change refers to continuous monotonic changes. However, other forms of monotonic change are covered by the definition of trend in (4.1), whether continuous or discontinuous. In the sense of (4.1), trend is a general concept that includes as special case shifts and monotonic step changes. On the one hand, the notion of trend is designed to capture long-term changes in the series (not due to the stochastic variation of the observations), and on the other hand, shifts reflect breakpoints in the series, caused for instance by abrupt events. In hydrometeorological applications, *seasonality* is also a different form of nonstationarity. A trend can be linear or not, and can also varies with respect to covariates, such as time or climate indices. Clearly time is the default covariate in most applications. In hydrology, one of the examples is for the estimation of design flows associated with a given date such as close to the end life of a hydraulic structure. Considering a climatic index is becoming as one of the popular and appropriate choices as a covariate for structure management, according to the index state the flooding risks can be re-estimated.

In the multivariate HFA context, a review and applications of nonparametric trend tests can be found in Chebana et al. (2013), which are described below. Since this testing step is not an end itself in HFA, in the case where a significant multivariate trend is detected, it is appropriate to consider nonstationary models (see Chapter 7).

4.2.1 Multivariate trend tests

In the univariate setting and for hydrological data, there exists a variety of methods to detect and assess trends (see e.g., Sonali & Kumar, 2013 for a review). These univariate approaches can be categorized into three groups: descriptive, nonparametric, and time series methods (Clement & Thas, 2009). This classification can be adapted and generalized to the multivariate setting. In HFA, the focus is more on the nonparametric methods in accordance with the aims of HFA as well as the main steps of the analysis.

In univariate statistical hydrology, one of the most widely employed nonparametric trend tests is the Mann–Kendall (MK) test. In addition, the Spearman rank order correlation test is another commonly used test. The World Meteorological Organization recommended both tests as standard nonparametric procedures when testing for trend. Indeed, they have the advantage of making very few assumptions and also have been shown to be more powerful than alternative procedures. Based on the additive model given in (4.1), trend tests are usually designed to test the null hypothesis H_0 of no trend: $T = 0$ against the general alternative hypothesis that there exists a monotonic trend T, that is H_1: $T \neq 0$ and there exists at least a component u such that $T^{(u)}$ is monotonic. In the following, we present a number of multivariate extensions of the above univariate tests. Note that the tests presented below were initially designed and developed for water quality applications, but later they have been considered in a number of multivariate HFA studies (Chebana & Ouarda, 2021; Karahacane et al., 2020).

4.2.1.1 Mann–Kendall type tests

For all the following tests, let $M^{(u)}$ be the univariate MK test statistic for the observed univariate time series $X_i^{(u)}, i = 1, \ldots, n$ and $u = 1, \ldots, d$. For a given component u, $M^{(u)}$ is given by:

$$M^{(u)} = \sum_{1 \le i < j \le n} \operatorname{sgn}\left(x_j^{(u)} - x_i^{(u)}\right) \tag{4.2}$$

where sgn(.) is the sign function $\operatorname{sgn}(x) = 1$ if $x > 0$; $= 0$ if $x = 0$; $= -1$ if $x < 0$. Under the null hypothesis H_0, that there is no significant monotonic trend, the statistic $M^{(u)}$ is asymptotically normally distributed with zero mean and approximated variance:

$$\operatorname{var}\left(M^{(u)}\right) = n(n-1)(2n+5)/18 \tag{4.3}$$

Hence, under H_0, the vector $M = \left(M^{(1)}, \ldots, M^{(d)}\right)'$ is asymptotically d-dimensional normal with zero mean and covariance matrix $C_M = (c_{u,v})_{u,v=1,\ldots,d}$ with $c_{u,v} = \operatorname{cov}\left(M^{(u)}, M^{(v)}\right)$. A consistent estimator of each covariance term $c_{u,v}$ is given by

$$\hat{c}_{u,v} = \frac{t_{u,v} + r_{u,v}}{3} \text{ for } u \neq v \tag{4.4}$$

where

$$t_{u,v} = \sum_{1 \le i < j \le n} \operatorname{sgn}\left(\left(x_j^{(u)} - x_i^{(u)}\right)\left(x_j^{(v)} - x_i^{(v)}\right)\right) \text{ and}$$

$$r_{u,v} = \sum_{i,j,k=1}^{n} \operatorname{sgn}\left(\left(x_k^{(u)} - x_j^{(u)}\right)\left(x_k^{(v)} - x_i^{(v)}\right)\right) \tag{4.5}$$

Here and in the following tests, H_0 is rejected if the value of the corresponding statistic is larger than a critical threshold related to the distribution of the underlying statistics. Alternatively, H_0 is rejected if the p-value is smaller than a given nominal value α (usually $\alpha = 5\%$).

4.2.1.2 *The covariance inversion test*

Let C_M^{-1} be the inverse matrix of C_M, when it exists, or a generalized inverse of C_M otherwise. The test statistic of the covariance inversion test (CIT) is given by:

$$D = M' C_M^{-1} M \tag{4.6}$$

It is has a $\chi^2(q)$ asymptotic distribution under H_0, where q is the rank of C_M with $1 \le q \le d$.

4.2.1.3 *The covariance sum test*

This test is also called the seasonal MK test. It falls in the multivariate setting since it was applied in a model assuming independent seasons, where each season is considered as a variable. Then, it was adapted to account for serial dependence. Therefore, the modified version can be employed to detect trends in multivariate correlated data. The test statistic is

$$H = \sum_{u=1}^{d} M^{(u)} \tag{4.7}$$

Under the null hypothesis, the statistic H is asymptotically centered normal, with variance:

$$\mathrm{var}(H) = \sum_{u=1}^{d} \mathrm{var}(M^{(u)}) + 2 \sum_{v=1,u=1}^{d,v-1} c_{u,v} \tag{4.8}$$

where $c_{u,v} = \mathrm{cov}\left(M^{(u)}, M^{(v)}\right)$ with an estimator given in (4.4).

4.2.1.4 *The covariance eigenvalue test*

This test was proposed in order to avoid matrix inversion as in (4.6) for the CIT test. The corresponding statistic is expressed as

$$L = M'M = \sum_{u=1}^{d} \left(M^{(u)}\right)^2 \tag{4.9}$$

Under H_0, given that the components $M^{(u)}$ are asymptotically normally distributed, if they are independent, then the statistic L is asymptotically $\sigma^2\chi^2(q)$-distributed. Recall that q is the rank of the covariance matrix C_M as in CIT. Nevertheless, the statistics $M^{(u)}, u = 1, \ldots, d$ could be dependent in general, and corresponding distribution of L cannot be obtained directly or in an exact form. Alternatively, it can be approximated. To this end, it is shown that the distribution of L is equal to the distribution of $L^* = \sum_{u=1}^{d} \lambda_u Z_u^2$ where λ_u are the eigenvalues of the covariance matrix C_M and Z_u are independent standard normally distributed random variables. In the case where all the λ_u are equal, that is, $\lambda_u = \lambda$ for all $u = 1,\ldots, d$, then both L^* and L are $\lambda\chi^2(d)$-distributed. If it is not the case, then a three-parameter Gamma distribution can approximate the distribution of L^*. The parameters associated to the three-parameter Gamma distribution of L^* are estimated as follows:

$$\alpha \approx \frac{\left(\sum_{u=1}^{d} \lambda_u^2\right)^3}{2\left(\sum_{u=1}^{d} \lambda_u^3\right)^2}, \quad \beta \approx \frac{2\sum_{u=1}^{d} \lambda_u^3}{\sum_{u=1}^{d} \lambda_u^2} \quad \text{and} \quad \gamma \approx \sum_{u=1}^{d} \lambda_u - \frac{\left(\sum_{u=1}^{d} \lambda_u^2\right)^2}{\sum_{u=1}^{d} \lambda_u^3} \tag{4.10}$$

4.2.1.5 *Spearman's rho type tests*

The Spearman-based tests are an interesting alternative to multivariate MK-based tests. Indeed, similarly to the MK tests, multivariate extensions based on the univariate Spearman statistic are available. The univariate version for the uth component of the observed data is given by

$$S^{(u)} = \sum_{i=1}^{n} \left(i - \frac{n+1}{2}\right)\left(\text{rank}\left(x_i^{(u)}\right) - \frac{n+1}{2}\right) \quad u = 1,\ldots,d \tag{4.11}$$

where $\text{rank}\left(x_i^{(u)}\right)$ is the rank of $x_i^{(u)}$ in the series of observations $x_1^{(u)},\ldots,x_n^{(u)}$. The reader is referred to Khaliq et al. (2009) for more details regarding the univariate Spearman trend test. Let $S = \left(S^{(1)},\ldots,S^{(d)}\right)'$ be the vector of the univariate Spearman statistics for each component. The associated covariance matrix $C_S = \left(c_{u,v}\right)_{u,v=1,\ldots,d}$ with elements $c_{u,v} = \text{cov}\left(S^{(u)}, S^{(v)}\right)$ are consistently estimated by

$$\hat{c}_{u,v} = \frac{n(n+1)}{12}\sum_{i=1}^{n}\left(\text{rank}\left(x_i^{(u)}\right) - \frac{n+1}{2}\right)\left(\text{rank}\left(x_i^{(v)}\right) - \frac{n+1}{2}\right) \tag{4.12}$$

The *Spearman-based CIT* statistic is given by:

$$B = S'C_S^{-1}S \tag{4.13}$$

Under the null hypothesis, as for the MK-based CIT (4.6), B is asymptotically $\chi^2(q)$-distributed, where $1 \le q \le d$ denotes the rank of the matrix C_S.

The equivalent test statistic to the MK-based covariance sum test (CST) (4.7) using the Spearman statistic can be expressed as

$$P = \sum_{u=1}^{d} S^{(u)} \tag{4.14}$$

This test takes into account the dependence between components and hence it is considered as a generalization of the Page test, which requires independent components.

In analogy with (4.9), a covariance eigenvalue test (CET) based on the Spearman statistic can be constructed and it is denoted R. To avoid repetition, explicit and full formulation of this test can be found in Chebana et al. (2013).

Table 4.1 gives an overview of the previously presented tests covering their expressions, asymptotic distributions, as well as the comparisons described below with more details.

4.2.2 Performance evaluation

It is important to emphasize that the following performance evaluations and comparisons were presented in the literature dealing essentially with water quality variables. Table 4.1 summarizes these results. The CIT could have relatively poor power for small sample sizes and is shown to be very conservative with respect to type I errors, that is, a large difference between the nominal and the empirical significance levels. For sample sizes around $n = 10$, the CIT power is very low. Even for larger samples, its power is lower than the power of CST and CET.

CST has better power than CIT and CET, and it performs well even for small sample sizes. In addition, CST provides a better estimate of type I errors. Nevertheless, the power of CST is negatively affected in the case where the univariate MK statistics $M^{(u)}, u = 1, \ldots, d$ had not the same sign, meaning that trends in different directions would cancel or reduce each other. Furthermore, this test was initially designed for seasonal data where each variable is associated with a season. One of the limitations of the CST test, in the multivariate setting, is that the trend magnitudes in the different variables is not necessary comparable. Hence, the additive form of the CST might be confusing.

An overall relatively good power is associated with CET which makes it as an interesting alternative. However, regarding to type I error estimation, the CET is relatively conservative. For large samples, the powers of CET and CIT become very similar. In this situation, CIT might be recommended due to its advantages including it is equally efficient, is easier to compute, and avoids estimation of parameters to approximate its distribution. In addition, if the components of the data are not correlated, then the CIT power increases compared to that of the original CIT as well as the CET with diagonal covariance matrix.

The Spearman-based multivariate tests, based on simulations, have similar results to those of MK-based tests. In this category, the CIT had low power for small sample sizes relatively to the dimension, and

TABLE 4.1 Overview of the presented multivariate trend tests.

	Name and expression	Asymptotic distribution under H_0	Comparisons: Within the same class	Comparisons: Overall
MK tests	Covariance Inversion $D = M'C_M^{-1}M$	$\chi^2(q)$, $q = \text{rank}(C_M)$, $1 \le q \le d$	For small n, D may have relatively poor power; with respect to type I errors, D is very conservative. D is basically powerless when n is around 10. For larger n, D has power less than the one of H and L. L has generally relatively good power and can be considered as an interesting alternative.	In the univariate setting, the powers of the MK- and the Spearman-based tests *are almost the same.* Generally, H and P do not show significant differences. P slightly outperforms H in the situation of data with no ties or missing values, and independent component variables.
	Covariance sum $H = \sum_{u=1}^{d} M^{(u)}$	Normal with zero mean and variance matrix given in (4.12)	The power of H could be affected if the univariate MK had not the same sign, H has better power than both L and D. With respect to type I errors, L is relatively conservative.	
	Covariance eigenvalue $L = \sum_{u=1}^{d} (M^{(u)})^2$	If $M^{(u)}$ are independent: $\sigma^2\chi^2(q)$, $q = \text{rank}(C_M)$, else approximated to three-parameter Gamma	For large n, the power gain of L over D became insignificant. In the absence of correlation between components, the power of the modified D is higher than both the ones of D as the modified L.	
Spearman tests	Covariance inversion $B = S'C_S^{-1}S$	$\chi^2(q)$, $q = \text{rank}(C_S)$, $1 \le q \le d$	R performed mostly better than B. B has low power for small n relatively to d, and relatively conservative with respect to type I errors.	
	Covariance sum $P = \sum_{u=1}^{d} S^{(u)}$	Similar to that of H		
	Covariance eigenvalue $R = \sum_{u=1}^{d} (S^{(u)})^2$	Similar to that of L		

Notations: n is the sample size and d is the dimension or the number of components.

seemed to be relatively conservative regarding to type I error estimation. The CET performed mostly better than the CIT.

When comparing the two categories, the MK- and the Spearman-based tests, the available results are in the univariate setting. Both test categories have almost identical powers. In the multivariate setting, it is expected to obtain similar results. Indeed, in the absence of an overall comparison, this statement is supported by a short simulation, which was conducted for the CST tests. In a quasi-ideal situation where there is no ties and no missing values in the data, and the component variables are independent, the Spearman-based CST outperforms slightly its MK-based counterpart. However, a more extensive comparison study is required in order to obtain more general results in the multivariate setting.

4.2.3 Further discussion

The above tests have been extended or generalized to nonstandard situations. For instance, extensions of the MK- and Spearman-based CIT tests are proposed to deal with the case of multivariate data series with missing values (Alvo & Park, 2002). In addition, in order to test series other than annual ones, for example, monthly series, the partial MK tests are developed. Incorporating a correction for covariates improves the original univariate and multivariate MK trend tests (El-Shaarawi, 1993). The performance of partial MK tests has been evaluated by Libiseller and Grimvall (2002) for both annual and seasonal data. In order to obtain a complete and accurate portrait of the multivariate trend analysis, it is recommended to jointly consider univariate and multivariate tests.

The magnitude of a trend is also of interest after a trend has been detected. Usually, a trend can be present for a long time series. Short-term trends have been studied in the univariate setting but not in multivariate one yet. In the latter, it could be more complex since there are different possibilities depending on the number of components (variables) where a trend is present, the magnitude of each trend as well as their directions (increasing or decreasing).

The MK- and Spearman-based tests, by their construction, are not able to detect non-monotonic trends. Even though the monotonic trends are realistic and occur in a majority of situations, other trend behaviors can be present, where they can be treated with semi-parametric or parametric methods (the reader is referred, for instance, to Clement & Thas, 2009).

4.2.3.1 Example

The previous trend tests are applied on data of the Moisie station located in the province of Quebec, Canada (code: 02UC002, Lat: 50.35, Long: −66.19). The length of the data series is $n = 35$ where data are available from 1966 to 2004 with three missing values. We considered three flood variables, that is, peak flow (*Q*), volume (*V*), and duration (*D*), which led to three pairs (*Q*, *V*), (*Q*, *D*), and (*D*, *V*).

Table 4.2 summarizes the statistics, the thresholds, and the *p*-values of the tests used in the univariate case. It shows that *Q* and *V* are not stationary, and there is a trend in the observations while *D* is found to be stationary. This can be seen in Fig. 4.1, which represents the corresponding time series.

Table 4.3 presents the results of the bivariate trend tests for the three pairs. For (*Q*, *V*), all the tests reject the null hypothesis and confirm that

TABLE 4.2 The univariate stationary tests for the Moisie station.

	Mann–Kendall test			Spearman test		
Variable	**|*K*|**	**Threshold**	***p*-value**	**|*S*|**	**Threshold**	***p*-value**
Q	**3.32**	**1.96**	**8.88e-04**	**4.10**	**2.03**	**2.59e-04**
V	**2.78**	**1.96**	**0.0054**	**3.59**	**2.03**	**0.0011**
D	0.60	1.96	0.55	0.25	2.03	0.80

Bold character indicates the presence of a trend at $\alpha = 0.05$.

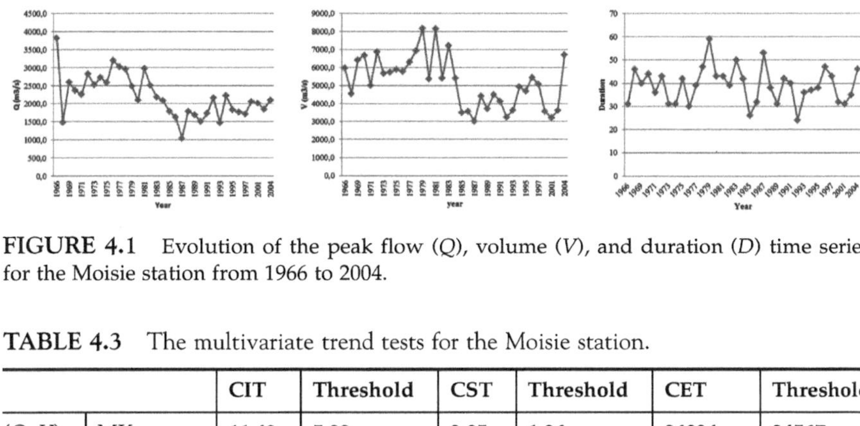

FIGURE 4.1 Evolution of the peak flow (*Q*), volume (*V*), and duration (*D*) time series for the Moisie station from 1966 to 2004.

TABLE 4.3 The multivariate trend tests for the Moisie station.

		CIT	Threshold	CST	Threshold	CET	Threshold
(*Q*, *V*)	MK	**11.43**	**5.99**	**2.05**	**1.96**	**94034**	**34567**
	Spearman	**12.21**	**5.99**	**2.17**	**1.96**	**7.8e + 6**	**2.6e + 6**
(*Q*, *D*)	MK	**11.88**	**5.99**	1.65	1.96	**57074**	**29655**
	Spearman	**11.71**	**5.99**	1.50	1.96	**4.3e + 6**	**2.2e + 6**
(*D*, *V*)	MK	**08.59**	**5.99**	1.21	1.96	**40658**	**32026**
	Spearman	**11.65**	**5.99**	1.15	1.96	**3.6e + 6**	**2.4e + 6**

Bold character indicates the presence of a trend at $\alpha = 0.05$.

there is a trend in the observations. For (Q, D) and (D, V), the null hypothesis is rejected with the CIT and the CET tests but not with the CST test. According to Chebana et al. (2013), the CET test is recommended for hydrologic data. Therefore, the three pairs can be considered as not stationary.

4.3 Homogeneity

In HFA, a series of observations is considered as homogenous if the whole series follows the same underlying distribution, which implies invariance of the corresponding statistical characteristics (e.g., mean, variance, symmetry). Therefore, it means that there is no mix of samples drawn from different populations. It is important to note that in this definition, time is not involved meaning that for a sample to be homogenous, the chronological order of the data is not required (Ben Nasr & Chebana, 2019; Rao et al., 2003). Several forms of departure from homogeneity can be encountered, in particular change-point (or shift).

According to the literature, either in statistics or hydroclimatology, detecting abrupt shifts is a common way to deal with homogeneity, in both univariate and multivariate contexts. In fact, homogeneity and shift are two related problems. As illustrated in Fig. 4.2, in the case of homogeneity, two or more pre-specified subgroups of data need to be tested (the chronological sequence of the data has no importance). However, when dealing with a shift, the appropriate subgroups of the data are unknown and need to be identified (the chronological sequence is required). Homogeneity testing can be treated as a shift testing by considering a preliminary step, such as classification or clustering. In addition, the shift detection and testing is largely studied in the literature. Based on these two elements, in the following, homogeneity is treated as a shift detection problem. To deal with the latter, a number of parametric and nonparametric tests have been proposed. In HFA, as discussed above, it is more appropriate to consider nonparametric approaches since they do not require prior choice of the underlying

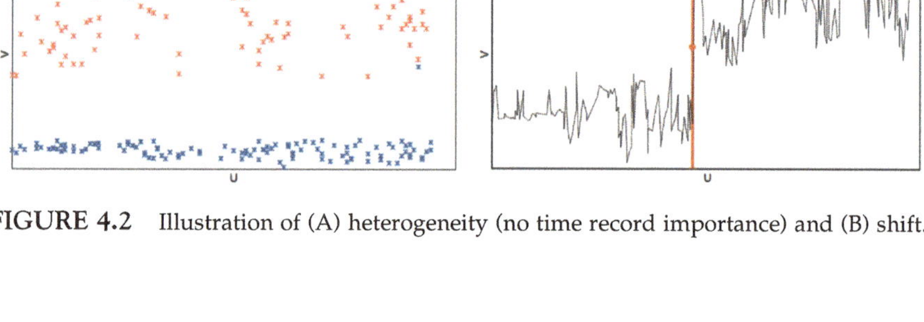

FIGURE 4.2 Illustration of (A) heterogeneity (no time record importance) and (B) shift.

distributions (margins or joint distributions). Santhosh and Srinivas (2013) showed a comparison between parametric and nonparametric tests in HFA.

Most of the univariate nonparametric tests for shifts in location consist in comparisons of the mean or the median, although there also exist tests comparing the whole empirical distribution function. The Wilcoxon rank-sum test, the equivalent Mann–Whitney U test, and the Kruskal-Wallis test are the classical univariate nonparametric tests. These tests have been found to be most effective among the available nonparametric tests. Other shift tests used for univariate hydrologic data include the Terry test, the Cramér–von Mises test, the Jonckheere test, or CUSUM tests. For an overview of the available methods to detect shifts in univariate data, the reader is referred to Kundzewicz and Robson (2004). The most frequently used and most powerful parametric competitor in both the univariate and the multivariate settings is Hotelling's T^2 test, which is only applicable under normality. In the statistical or hydrometeorological literature, a large number of methods have been developed and applied to identify the date of a potential shift and to check the significance of the associated shift. An important portion of these methods employs statistical hypothesis testing to detect changes in slopes or intercept of linear regression models (Lund & Reeves, 2002).

Unlike the univariate tests, the application of multivariate tests for shifts in hydrology and climatology has not yet received much attention (e.g., Chebana et al., 2017). Homogeneity of multivariate hydrological series can be affected at marginal distributions and/or the dependence structure (e.g., Ben Nasr & Chebana, 2019; Xiong et al., 2015). In the multivariate hydrological context, Chebana et al. (2017) presented and evaluated a general class of tests, based on statistical depth function (see Appendix). These tests are designed to detect an overall shift in the series and not focusing on any of its components (each of the margins or the dependence structure). Testing for homogeneity in each of the margins is part of the univariate HFA. However, to treat homogeneity in the dependence structure, recently multivariate tests have been introduced (e.g., Quessy et al., 2013; Xiong et al., 2015). Nonetheless, these tests have some drawbacks, specifically for HFA, which are presented in Ben Nasr and Chebana (2019).

In hydrological data series, the commonly used form of testing the homogeneity is about shift detection on location and/or scale parameters. This formulation relies on the sequential time occurrences of the events. However, especially in HFA, heterogeneity can be present without importance to the sequential aspect. In statistics, the latter means that the whole distribution may change abruptly and not only some of its features (e.g., Serinaldi et al., 2018). In addition, the shift change testing can be seen as the *two-sample location problem* where the positions of the shifts may be known or not. In practice, no prior information is available about possible shifts for a given series of hydrologic observations. Even though formal techniques are available, visualization can be a simple and useful tool to identify dates of changes. The tests presented below are based on the assumption that the shift date is already known.

Let $(x_i)_{i=1,\ldots,n}$ be a given series of size n and dimension d and $1<s<n$ be a possible shift. The existence of s means the series is divided into two subsamples:

$$(y_1,\ldots,y_s) = (x_1,\ldots,x_s) \text{ and } (z_1,\ldots,z_m) = (x_{s+1},\ldots,x_n) \tag{4.15}$$

respecctively with sizes s and $m = n\text{-}s$, and distributions G_1 and G_2. In this context, the distributions G_1 and G_2 have the same form, except for location, that is, $G_1(x) = G_2(x + \delta)$ for all $x \in R^d$ where $\delta \in R^d$ is a constant location vector. Therefore, the null and alternative hypotheses are respectively:

$$H_0{:}\delta = 0 \text{ i.e. there is no location shift} \tag{4.16}$$

$$H_1{:}\delta \neq 0 \text{ i.e. there are two different subsamples (in at least one component of } \delta) \tag{4.17}$$

4.3.1 Multivariate shift detection tests

The majority of the available shift tests in the literature are designed to detect shift in location. All the tests considered and presented below are based on data depth function, except the C-test. Recall that a depth function is a statistical notion aiming to ranking multivariate data, where a brief description of depth functions is given in Appendix. In this section, the following depth functions are considered: Mahalanobis (MD), simplicial (SD), and half-space (or Tukey, TD). Even though conceptually each depth-based test can be constructed using any depth function, usually a given test is based on a specific depth function in its definition and the study of its properties. Table 4.4 summarizes the following presented tests.

4.3.1.1 *The Cramér test*

This test is a two-sample test, initially developed for the univariate case. It is more appropriate to detect shifts in location. The corresponding test statistic is given by

$$C = \frac{sm}{s+m}\left[\frac{1}{sm}\sum_{i=1}^{s}\sum_{j=1}^{m}\left\|y_i - z_i\right\| - \frac{1}{2s^2}\sum_{i,j=1}^{s}\left\|y_i - y_j\right\| - \frac{1}{2m^2}\sum_{i,j=1}^{m}\left\|z_i - z_j\right\|\right] \tag{4.18}$$

TABLE 4.4 Summary of the presented tests and comparison of their performances.

<table>
<tr><th rowspan="2"></th><th rowspan="2">Designed to detect shift in</th><th rowspan="2">p-value evaluation</th><th rowspan="2">Used depth functions</th><th colspan="3">Comparisons</th></tr>
<tr><th>Normal samples</th><th>Non-normal samples</th><th>Hydrological context</th></tr>
<tr><td>M-test</td><td>Location</td><td>Permutation</td><td>Simplicial[a]; Mahalanobis; Half-space</td><td rowspan="2">The powers of M-test, T-test, and Hotteling tests are comparable</td><td rowspan="2">The M-test outperformed the T-test and both are more powerful than the Hotelling test</td><td rowspan="6">The QIA, QIB, and Z tests may be problematic when no change occurs in the marginal variables. The C-test power is sensitive to the magnitude of variables. It can be recommended to use the M-test (Mahalanobis), T-test (half-space or simplicial), W test (half-space or simplicial), or Z test (Mahalanobis)</td></tr>
<tr><td>T- test</td><td>Location</td><td>Permutation</td><td>Simplicial[a]; Mahalanobis</td></tr>
<tr><td>Q-test</td><td>Location and/or positive scale</td><td>Bootstrap or asymptotic</td><td>If p-value found asymptotically: Mahalanobis[a] or Half-Space If bootstrap is the p-value evaluation: Half-Space or Simplicial</td><td>The performances of the Q- and Hotelling T tests are similar</td><td>The Q-test outperformed the Hotelling one</td></tr>
<tr><td>Z-test</td><td>Multiple location and/or scale</td><td>Asymptotic</td><td>Half-Space ; Mahalanobis</td><td colspan="2">NA</td></tr>
<tr><td>W-test</td><td>Location</td><td>Critical thresholds</td><td>Half-Space ; Simplicial[a]; Mahalanobis</td><td colspan="2">NA</td></tr>
<tr><td>C-test</td><td>Location and/or scale</td><td>Bootstrap</td><td>NA</td><td colspan="2">The C-test performs similar to Hotelling test</td></tr>
</table>

[a] *With which the test was originally developed.*

where $\|y_i - z_j\|$ is the Euclidian distance between ith observation of first subsample and jth observation of second subsample. The null hypothesis H_0 is rejected for large value of C. The critical value, depending on the sample sizes and the fixed significance level, is determined via bootstrapping. The Cramér test (C-test) statistic is consistent and orthogonally invariant. Baringhaus and Franz (2004) presented extended versions of the C-test using a generalized concept of multivariate distances instead of Euclidean distances, as well as details about the asymptotic properties of this class of tests.

4.3.1.2 *The M-test*

This test is based on the fact that the deepest point of a distribution is a location parameter, which generally true as seen in Chapter 3 (see multivariate medians). Accordingly, if the distributions G_1 and G_2 of the two subsamples are identical, then they would share the same deepest point. In other words, their respective deepest point θ_{G_1} and θ_{G_2} should be equal $\theta_{G_1} = \theta_{G_2}$ and $D_{G_2}(\theta_{G_1}) = D_{G_1}(\theta_{G_2})$ for a given depth function D. However, if there is an important change in location, θ_{G_1} and θ_{G_2} would be different and θ_{G_2} is located far away from the subsample with the distribution G_1 for which the depth value $D_{G_1}(\theta_{G_2})$ with respect to G_1 is smaller. Based on this idea, the corresponding statistic is given by

$$M = \min\{D_{G_2}(\theta_{G_1}), D_{G_1}(\theta_{G_2})\} \tag{4.19}$$

The null hypothesis H_0 can be accepted for large value of M. Fisher's permutation was considered in order to approximate the corresponding p-value (see Appendix). In general, simplicial and half-space depth functions were used to define M, and for elliptical desitribution Mahalanobis depth was suggested.

4.3.1.3 *The T-test*

The statistic of this test compares the location of two subsamples based on the DD-plot (for Depth-Depth). In this context, the DD-plot represents the depth values of two subsamples on each of its axes. Similar to QQ-plot, the DD-plot is a diagonal line that passes by the origin in the case where the two subsamples follow exactly the same distribution. However, in the case of a location change, the DD-plot has a form of leaf with a tip pointing toward the origin. The more the location change is high, the more the tip becomes close to the origin. Hence, the distance between the tip and the origin of the DD-plot allows to define the test statistic. Formally, it is given by

$$T = \frac{D_{G_1}(x_{\min}) + D_{G_2}(x_{\min})}{2} \tag{4.20}$$

where the point $x_{\min}$ of Ω such that $|D_{G_1}(x_{\min}) - D_{G_2}(x_{\min})| = \min_{x \in \Omega} |D_{G_1}(x) - D_{G_2}(x)|$ and the set Ω is

$$\Omega = \{x_i | i \in \{1, \ldots, n\}, \text{ there is no } x_j : D_{G_1}(x_j) \geq D_{G_1}(x_i) \text{ and } D_{G_2}(x_j) \geq D_{G_2}(x_i)\} \tag{4.21}$$

If there is more than one point $x_{\min}$, the mean of their coordinates is taken.

Similar to the M-test, the simplicial depth function was used for the T-test, but the Mahalanobis and half-space depths can also be used. The asymptotic distribution corresponding to T is not straightforward and not yet available. However, a permutation method is employed to approximate the associated p-values (see Appendix).

4.3.1.4 The Wilcoxon test

Similar to the M-test, the Wilcoxon test (W-test) relies on the idea that the medians of the two subsamples must be similar under null hypothesis. The W-test statistic computation is based on the difference of each component $d_{ij}^{(u)} = z_i^{(u)} - y_j^{(u)}, u = 1, \ldots, d;\ i = 1, \ldots, s;\ j = 1, \ldots, m$ to constitute the vector $d_{ij} = \left(d_{ij}^{(1)}, \ldots, d_{ij}^{(d)}\right)$. The test statistic is given by

$$W = \frac{D_F(\mathbf{0})}{\max\limits_{i=1,\ldots,s;j=1,\ldots,m} D_F(d_{ij})} \tag{4.22}$$

where F is the distribution of the set of vectors d_{ij} and D_F is the half-space depth function corresponding to F. Under the null hypothesis, $W = 1$, whereas under the alternative hypothesis, $W < 1$. The asymptotic distribution of W is unknown, but some critical values C_α are available such that $C_\alpha = 1 - a_\alpha / \sqrt{\min(s, m)}$ where $a_\alpha = 2.1668, 2.0215, 1.8556, 1.6338$ respectively for $\alpha = 0.01, 0.025, 0.05, 0.10$. The null hypothesis is rejected when W is below C_α.

4.3.1.5 The quality index test

This test is able to detect location shift and/or positive scale shift. The corresponding statistic is given by

$$Q_a = \frac{1}{n} \sum_{i=1}^{m} \#\{y \in \{y_1, \ldots, y_s\} : D_G(y) \leq D_G(z_i)\} \tag{4.23}$$

In the absence of a shift (under H_0), $Q_a = 0.5$, otherwise $Q_a < 0.5$ if there is a shift in location. Note that Mahalanobis depth function was used. In addition, under some regularity conditions, the asymptotic distribution of Q_a is normal when considering Mahalanobis, half-space, or projection depths. Hence, the p-value can be evaluated either asymptotically (denotes QIA test) or by bootstrap (denoted QIB test). Although this test can detect shifts in both location and scale, it is more sensitive to changes in scale than to changes in location.

4.3.1.6 The Zhang test

The Zhang test (Z-test) is similar to the quality index test where the statistic Q_a is transformed leading to:

$$Z = \frac{6\,s\,m}{n}(Q_a - 0.5)^2 \qquad (4.24)$$

Similar to the quality index test, the Mahalanobis depth function was used to define Z, and half-space and projection depth functions are valid alternatives. Under the null hypothesis and with these depth functions, Z is asymptotically a sum of independent chi-square distributions. This asymptotic result allows to obtain the corresponding p-value.

4.3.2 Comparisons and other approaches

Chebana et al. (2017) compared the above tests within the hydrological context based on a simulation study. Results showed that the power increases with the shift amplitude and the sample size. The QIA, QIB, and Z tests may be problematic when no change occurs in the marginal variables. For low shift amplitude (e.g., 10%), the considered tests do not perform well irrespective of the sample length. The C-test power is sensitive to the magnitude of variables. The following tests have been recommended for these comparisons, along with the nature of hydrological data: M-test (with Mahalanobis depth), T-test (with half-space or simplicial depth), W-test (with Mahalanobis depth), or Z-test (with Mahalanobis depth). Note that a number of other studies compared some of these tests as well. However, these comparisons do not necessary take into account the context of hydrological applications, where sample sizes are generally short, and the distributions are not normal but of extremes among other considerations.

A number of other multivariate tests are available in the literature to test two-sample hypotheses, for instance, tests based on the Oja median, interdirections, multivariate spatial sign and rank methods and their generalizations. Based upon a similar idea as the C-test, Fernández et al. (2008) suggested a test using the empirical characteristic function of the data. Mathur (2009) proposed a bivariate test for a variety of distributions which displayed good power in comparison to other tests.

Shift detection is not always based on hypothesis testing. For instance, curve fitting methods are considered such as the gray relational method (e.g., Wong et al., 2006). The latter is used for single shift detection in streamflow data series. Beaulieu et al. (2007) reviewed shift detection and correction methodologies in hydroclimatology.

4.3.3 Example

All the homogeneity tests explained above are applied for the Moisie station (02UC002). First, the univariate W-test is applied to each of the flood variables where the subsamples are obtained either the shift year is 1987 or 1983, see below for more description. In both situations, it can be seen that both Q and V series are not homogeneous which not the case for D series (Table 4.5).

The above multivariate tests are applied to the three pairs (Q, V), (Q, D), and (D, V). It is divided into two subsamples to check the homogeneity of the bivariate series. As shown in Fig. 4.3, the evolution of these series shows that 1987 seems to be an exceptional year that divides the

TABLE 4.5 Results of the univariate Wilcoxon test homogeneity test for the Moisie station.

	Variable	$\|W\|$	Threshold	p-value
Moisie station (year 1987)	Q	**2.95**	1.96	**0.0032**
	V	**2.71**	1.96	**0.0067**
	D	0.73	1.96	0.47
Moisie station (year 1983)	Q	**4.11**	1.96	**0.000040**
	V	**4.01**	1.96	**0.000061**
	D	0.84	1.96	0.40

Bold character indicates the presence of a shift at $\alpha = 0.05$.

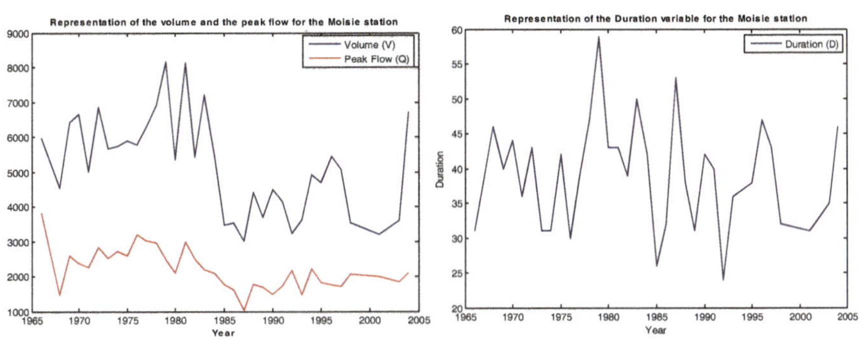

FIGURE 4.3 Peak flow, volume, and duration time series of the Moisie station.

data into two subsamples. Hence, we considered two subsamples according to this year. The first subsample has a length $s = 21$, while the second one has a length $m = 35\text{-}s = 14$. Similarly for the year 1983. Table 4.6 summarizes the results of the tests listed above for the Moisie station.

Bold values mean significant heterogeneity between the two subsamples (detected shift). The resultants are presented as statistics (p-value or decision for W-test). For all the tests presented in Table 4.6, the p-value is calculated using a bootstrap procedure ($N = 10000$) except QIA which is asymptotic (see also Appendix: p-value computation).

The results of all the tests presented in Table 4.6 for the Moisie station show that the two subsamples considered for (Q, V), (Q, D), and (D, V) data are not homogeneous when considering 1983. However, we have some exceptions with 1987 in particular with QIA, QIB and Z-tests which are not the recommended ones (as discussed above). In fact, large values of the statistic C of the Cramér test are obtained that are associated with low p-values, which mean that the distance between the observations of the two subsamples is large so the two subsamples are different. For the M and T tests, the simplicial depth is used since it can be used with any distribution. For small values of the statistic M, the null hypothesis is rejected so the two-subsamples are different. For the T test, the p-values found are very low so the null hypothesis is rejected.

For the W test, if the statistic $W < 1$, the null hypothesis is then rejected. Therefore, the decision is 0 if there is no change in data and the homogeneity is accepted. Otherwise, the decision is 1, if there is a change in the data set and the homogeneity is not satisfied which is the case for the Moisie station. For the quality index test, at significance level of 5% the homogeneity hypothesis is rejected. Z test confirms that there is a shift in location since p-values are considerably less than 0.05 for all the tested data sets.

4.4 Serial independence

Serial dependence is often present and must be taken into account in statistical inference by testing it and, where appropriate, integrating it into the model to be selected. Serial dependence is the essential element in time series modeling, including copula modeling. On the other hand, it is important to mention that the finalities and aims of the time series as a statistical field are different from those of the HFA. Multivariate serial dependence tests have been proposed in the statistical literature but are not yet considered in multivariate HFA. In the univariate HFA, the Wald–Wolfowitz test is usually used to test for

TABLE 4.6 Results of the multivariate tests for the pairs of the Moisie station data.

		(Q, V)		(Q, D)		(D, V)	
Moisie (1987)		**Statistics**	***p*-value**	**Statistics**	***p*-value**	**Statistics**	***p*-value**
	C-test	**5488.45**	**<0.001**	**2269.57**	**<0.001**	**4854.35**	**<0.001**
	M-test	***0.079***	***<0.001***	***0.094***	***<0.001***	**0.22**	***<0.001***
	T-test	***0.69***	***0.028***	***0.53***	***<0.001***	***0.59***	***<0.001***
	W-test[a]	***0.50***	***0[a]***	***0.48***	***0[a]***	***0.56***	***0[a]***
	QIA test	0.37	0.10	0.39	0.14	0.40	0.16
	QIB test	0.37	0.15	0.39	0.21	0.40	0.23
	Z-test	0.80	0.37	0.56	0.45	0.49	0.48
Moisie (1983)							
	C-test	9849.72	<0.001	3887.41	<0.001	9030.82	<0.001
	M-test	*0.098*	*<0.001*	*0.081*	*<0.001*	*0.15*	*<0.001*
	T-test	*0.39*	*<0.001*	*0.30*	*<0.001*	*0.43*	*<0.001*
	W-test	*0.24*	*0[a]*	*0.24*	*0[a]*	*0.30*	*0[a]*
	QIA test	0.12	0.000052	0.16	0.00028	0.14	0.00013
	QIB-test	0.12	<0.001	0.16	0.001	0.14	<0.001
	Z-test	7.54	0.0058	5.94	0.015	6.66	0.0094

[a]*This is not a p-value but a decision as 0 (reject) or 1 (accept) of H0.*
Bold character indicates the presence of a shift at $\alpha = 0.05$. Italic character indicates recommended tests.

the (serial) independence [see e.g., Rao and Hamed (2000) for more details], even though a variety of other tests are available in the literature and considered in different applications. The serial independence testing, especially in HFA, is not as developed as for homogeneity and stationarity cases, especially in the multivariate setting. Even though Herwartz and Maxand (2018) reviewed the available tests for independence including multivariate ones, their study does not cover *serial* dependence tests.

In multivariate time series modeling, a natural first step is testing for multivariate serial independence. If serial dependence is detected, a multivariate model that takes into consideration both serial and cross dependence must be developed. Usually, to test the presence of serial dependence, one can estimate the autocorrelation and cross-correlation functions of the multivariate time series as well as Ljung–Box-type portmanteau tests and the multivariate version of Wald–Wolfowitz rank test. In the univariate case, some powerful tests of serial independence are based on the empirical distribution function. These approaches can be regarded as the serial analogs of the well-known test of independence proposed by Blum et al. (1961). However, for heavy-tailed observations, the latter are often too liberal (Hofert et al., 2018). Alternatively, Ghoudi et al. (2001) investigated serial independence tests based on a *Möbius decomposition*. Later, Genest and Rémillard (2004) studied a version of this decomposition based on the serial empirical copula to propose a test based on ranks. Based on a multivariate version of the Möbius decomposition, proposed by Beran et al (2007), Kojadinovic and Yan (2011) proposed a multivariate extension of the above test and also provided practical computation of the statistics and the corresponding p-values. This test is presented below and applied to a case study as an illustrative example.

4.4.1 Serial empirical copula test

Let $X_1, X_2, \ldots$ be a stationary (and ergodic) univariate sequence of continuous random variables and an integer $p > 1$. First, p-dimensional vectors should be formed of observations $Y_i = (X_i, \ldots, X_{i+p-1})$, $i \in \{1, \ldots, n\}$, where p is the embedding dimension or alternatively called the maximum lag p - 1. From the stationarity assumption, all the p-dimensional vectors Y_i have the same distribution with cumulative distribution function (c.d.f.) H. The latter can be expressed via a unique copula C: $[0, 1]^p \rightarrow [0, 1]$, such that $H(x_1, \ldots, x_p) = C[F(x_1), \ldots, F(x_p)]$, for $(x_1, \ldots, x_p) \in R^p$, where F is the common c.d.f. of each X_i (copulas are described in Chapter 5 where all the required details can be found). Since under serial independence of $X_1, X_2, \ldots$, the copula $C(u_1, \ldots, u_p)$ coincides with the independence copula $\prod_{k=1}^{p} u_k$, for $(u_1, \ldots, u_p) \in [0, 1]^p$, then the test statistics is derived from the process

$$\sqrt{n}\left\{ C_n^s(u_1, \ldots, u_p) - \prod_{k=1}^{p} u_k \right\}, \quad u_1, \ldots, u_p \in [0, 1]^p \tag{4.25}$$

where C_n^s is the serial version of the empirical copula calculated from $Y_1, Y_2, \ldots, Y_n$. Under serial independence of $X_1, X_2, \ldots$, the empirical process in (4.25) can be decomposed, using the *Möbius transform*, into a collection of subprocesses $\sqrt{n}M_A(C_n^s)$, where A is a set such that $A \subseteq \{1, \ldots, p\}$, $1 \in A, |A| > 1$. The collection of $\sqrt{n}M_A(C_n^s)$ converges jointly to tight-centered mutually independent Gaussian processes. The function M_A from $\ell^\infty[0,1]^p$ to $\ell^\infty[0,1]^p$ is given by $M_A(f)(x) = \sum_{B \subseteq A} (-1)^{|A|-|B|} f(x_B) - \prod_{k \in A\, B} f(x_{\{k\}})$, $x \in [0,1]^p$ where $\ell^\infty[0,1]^p$ is the space of all bounded real-valued functions on $[0,1]^p$.

Instead of a single Cramér–von Mises test statistic based on (4.25), $\sqrt{n}M_A(C_n^s)$ leads to consider a collection of statistics of the form

$$n \int_{[0,1]^p} \left[M_A\left(C_n^s\right)(u)\right]^2 du, \quad A \subseteq \{1, \ldots, p\}, \ 1 \in A, |A| > 1 \tag{4.26}$$

These statistics are asymptotically mutually independent under serial independence (the null hypothesis). This decomposition can be interpreted as each of these Cramér–von Mises statistics can be seen as focusing on a particular type of departure from serial independence.

As indicated above, Kojadinovic and Yan (2011) generalized the test proposed by Genest and Rémillard (2004) to the continuous multivariate time series setting as well as validating the corresponding bootstrap approach. To this end, consider a stationary (ergodic) sequence of d-dimensional continuous random vectors $\mathbf{X}_1, \mathbf{X}_2, \ldots$, the common c.d.f. of each $\mathbf{X}_i$ is denoted by F and the corresponding copula by C. A natural extension of the serial independence empirical copula process (4.25) to the multivariate time series setting is then

$$\sqrt{n}\left\{ C_n^s(u) - \prod_{j=1}^{p} C_n^s(u_{\{k\}}) \right\}, \quad u \in [0, 1]^{pd} \tag{4.27}$$

where

$$C_n^s(u) = \frac{1}{n} \sum_{i=1}^{n} \prod_{j=1}^{p} \prod_{k=1}^{d} 1\left[F_n^{(k)}\left(X_{i+j-1}^{(k)}\right) \le u_{\langle j \rangle}^{(k)}\right] = \frac{1}{n} \sum_{i=1}^{n} \prod_{j=1}^{p} \prod_{k=1}^{d} 1\left[R_{i+j-1}^{(k)} \le n' u_{\langle j \rangle}^{(k)}\right] \tag{4.28}$$

is the serial empirical copula computed from $\mathbf{Y}_1, \ldots, \mathbf{Y}_n$, the pd-dimensional random vectors $\mathbf{Y}_i = (\mathbf{X}_i, \ldots, \mathbf{X}_{i+p-1})$ with $\mathbf{X}_i = (X^{(1)}{}_i, \ldots, X^{(d)}{}_i)$,

$i \in [n]$, $[k]$ is the set $\{1,\ldots, k\}$ for any integer $k>0$ with $[0]=\varnothing$. Furthermore, let $p>1$ be an integer, $n' = n + p - 1$ and, for any $j \in [q]$, let $R^{(j)}{}_1, R^{(j)}{}_2, \ldots, R^{(j)}{}_{n'}$ be the ranks associated with the univariate sequence $X^{(j)}{}_1, X^{(j)}{}_2, \ldots, X^{(j)}{}_{n'}$. The empirical c.d.f. computed from the random vectors $\boldsymbol{X}_i, \ldots, \boldsymbol{X}_n$ is denoted by F_n, that is,

$$F_n(x) = \frac{1}{n'}\sum_{i=1}^{n'}\prod_{k=1}^{d} 1\left[X_i^{(k)} \leq x^{(k)}\right] = \frac{1}{n'}\sum_{i=1}^{n'} 1[X_i \leq x], \quad x \in R^d \tag{4.29}$$

where inequalities between vectors are component-wise. The corresponding marginal c.d.f.s are denoted by $F_n{}^{(1)}, F_n{}^{(2)}, \ldots, F_n{}^{(d)}$. The ranks $R^{(j)}{}_i$ are then related to the $X^{(j)}{}_i$ through $R^{(j)}{}_i = n'F^{(j)}{}_n(X^{(j)}{}_i)$, $i \in [n']$, $j \in [d]$. Given a set $B \subseteq [p]$ and $\mathbf{u} \in [0, 1]^{pd}$, let us define the vector $\mathbf{u}_B \in [0, 1]^{pd}$ *as*

$$u_B^{(i)} = \begin{cases} u^{(i)}, & \text{if } i \in \cup_{j \in B}\{(j-1)d+1, \ldots, jd\} \\ 1, & \text{otherwise.} \end{cases}$$

Given $\mathbf{u} \in [0, 1]^{pd}$ and $j \in [p]$, the vector $\mathrm{u}_{\langle j \rangle} \in [0,1]^d$ is defined by $u_{\langle j \rangle}^{(i)} = u^{(i+(j-1)d)}$, $i \in [d]$. The vector $\mathrm{u}_{\langle j \rangle}$ is a subvector of $\mathbf{u}$, whereas $\mathbf{u}_{\{j\}} \in [0, 1]^{pd}$.

Under some conditions, Kojadinovic and Yan (2011) established the asymptotic distribution of the serial empirical copula process given in (4.27). Based on this result, one can define the test statistics by measuring departure from serial independence using the corresponding Cramér–von Mises statistic leading to:

$$I_n = n\int_{[0,1]^{pd}}\left[C_n^s(\mathrm{u}) - \prod_{k=1}^{p} C_n^s(\mathrm{u}_{\{k\}})\right]^2 d\mathrm{u} \tag{4.30}$$

The asymptotic distribution of I_n is also obtained under serial independence of $\boldsymbol{X}_1, \boldsymbol{X}_2, \ldots$.

As in the univariate test by Genest and Rémillard (2004), in order to obtain potentially more powerful tests, a collection of 2^{p-1}-1 test statistics can be defined based on the *Möbius decomposition* of the process (4.2):

$$M_{A,n} = n\int_{[0,1]^{pd}}\left[M_A\left(C_n^s\right)(u)\right]^2 du, \quad A \in P_1 \tag{4.31}$$

where $P_1 = \{B \in P\colon 1 \in B\}$ and $P = \{B \subseteq [p]\colon |B| > 1\}$. If C has continuous partial derivatives, then, under serial independence of $\boldsymbol{X}_1, \boldsymbol{X}_2, \ldots$, the random vector $\{M_{A,n}\colon A \in P_1\}$ converges in distribution to a random vector.

As a result, we obtain the global Cramer–von Mises test I_n as well as a set of tests $M_{A,n}$, $A \in P_1$, each of which has its own p-value. Under serial independence, p-values of $M_{A,n}$, $A \in P_1$ are approximately independent and uniform on [0, 1]. Hence, these p-values can be combined in order to make a decision to accept or to reject the serial independence hypothesis. These p-values can be combined either à la Fisher or à la Tippet. Formulas of Kojadinovic & Yan (2011).

The selection of the embedding dimension p seems to be a common issue to all tests using it. To the best of our knowledge, there is no formal statistical inference procedures available in the literature for this purpose (Kojadinovic & Yan, 2011). However, Bücher et al. (2019) discussed this issue and recommend, even roughly, to take $p = 2$, 3, or 4. In addition, Genest and Rémillard (2004) suggested a graphical representation called a *dependogram* of the values of the observed test statistics. A dependogram is composed of vertical bars where one corresponds to each subset $A \in P_1$, with height represents the value of $M_{A,n}$.

In addition, a critical value is indicated on each bar, corresponding to the corrected significance level $1-(1-\alpha)^{1/(2^{p-1}-1)}$. If for a subset A, the bar exceeds the critical value, then it can be considered as composed of dependent vectors. For an illustration of the dependogram, see Fig. 4.1 in Kojadinovic and Yan (2011) where p is taken to be equal to 6. The authors also presented an interesting application section (section 7 in Kojanovitch & Yan, 2011).

4.4.2 Illustrative example

We test the cross dependence for the (Q, V), (Q, D), and (D, V) couples of the Moisie station (02UC002), the same data considered in the previous sections. The results are summarized in Table 4.7. For the three pairs, all the p-values are higher than 5%. Hence, the series can be considered as serially independent. This is expected since the data represent annual floods which are more likely to be serially independent.

The results of the pair (Q, V) are analyzed in Table 4.8 and Fig. 4.4. The embedding dimension is $p = 5$, which leads to the subsets A as in

TABLE 4.7 Summary of the serial dependence results for the Moisie station.

Pairs	Global Cramér-von Mises statistic I_n	p-value	p-value from Fisher's rule	p-value from Tippet's rule
(Q, V)	0.00196	0.117	0.177	0.352
(Q, D)	0.00032	0.276	0.139	0.266
(D, V)	0.00066	0.466	0.693	0.649

TABLE 4.8 The subset statistics, p-values, and critical values for the pair (Q,V) for the Moisie station.

Subset A	Statistic $M_{A,n}$	p-value	Critical value
{1,2}	0.01831	0.6908	0.06970
{1,3}	0.02194	0.4780	0.06970
{1,4}	0.02315	0.4060	0.06970
{1,5}	0.04104	0.0684	0.06970
{1,2,3}	0.00324	0.5569	0.00778
{1,2,4}	0.00334	0.5030	0.00778
{1,2,5}	0.00358	0.3831	0.00778
{1,3,4}	0.00328	0.5200	0.00778
{1,3,5}	0.00413	0.1703	0.00778
{1,4,5}	0.00445	0.1174	0.00778
{1,2,3,4}	0.00053	0.3941	0.00111
{1,2,3,5}	0.00058	0.2452	0.00111
{1,2,4,5}	0.00076	0.0395	0.00111
{1,3,4,5}	0.00062	0.1693	0.00111
{1,2,3,4,5}	0.00010	0.0904	0.00022

the first column in Table 4.8. Accordingly, there is no joint dependence between the subsets at a significance level of 5% since the p-values are higher than 5%. Also the statistics are very low compared to their corresponding critical values. Only one exception is for the subset $A =$ {1,2,4,5}, where the p-value is very slightly lower than 5%, which can also be seen from Fig. 4.4.

4.5 Complete illustrative example

In this example, we used the data of the Nottawasaga River near Baxter (station number 02ED003) in Ontario (Canada), which is considered in Chapters 2 and 3. For simplicity and readability, we focused on the pair (Q,V) as the most studied in flood multivariate HFA. The aim here is to provide an example where we checked all the three assumptions (stationarity, homogeneity and serial-independence) on the same dataset.

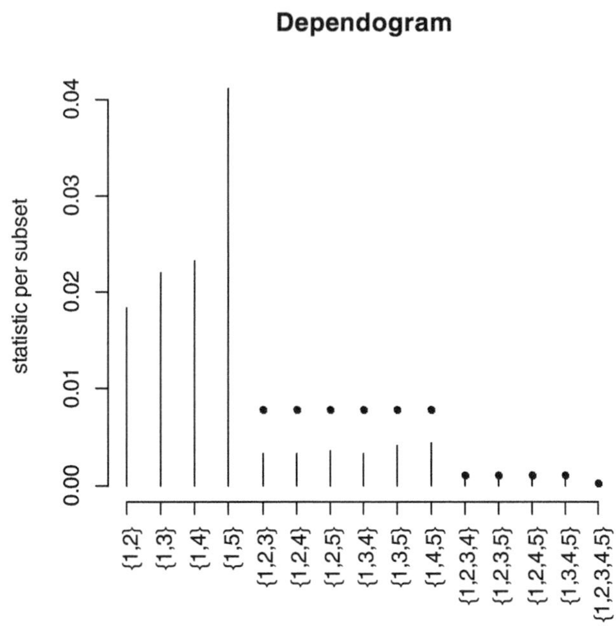

FIGURE 4.4 Dependogram of the subsets for the (Q, V) couple for the Moisie station. Points represent critical values (they are not shown for the first four subsets since they are high, see Table 4.8).

TABLE 4.9 The univariate and multivariate trend tests for the Nottawasaga River (station 02ED003).

Univariate tests							
	Mann–Kendall test				Spearman test		
Variable	$\|K\|$	**Threshold**	***p*-value**	$\|S\|$	**Threshold**	***p*-value**	
Q	−0.83	1.96	0.40	−0.74	2.03	0.47	
V	−1.28	1.96	0.20	−1.29	2.03	0.21	
Multivariate tests							
	CIT	**Threshold**		**CST**	**Threshold**	**CET**	**Threshold**
(Q, V)	MK	1.78	5.99	0.77	1.96	16349	43609
	Spearman	1.67	5.99	0.72	1.96	1387880	4.1013e + 06

First, from Table 4.9, we observe that the trend hypothesis is rejected by all the tests, in both univariate and multivariate frameworks.

The homogeneity assumption is plausible given the p-values in Table 4.10 by all the tests in the univariate and multivariate frameworks

TABLE 4.10 Results of the univariate and multivariate homogeneity tests for the Nottawasaga River (station 02ED003).

	Statistics	***p*-value**
Univariate for V		
Wilcoxon test	0.80 (threshold 1.96)	0.42
Univariate for Q		
Wilcoxon test	0.24 (threshold 1.96)	0.81
Multivariate for (Q, V)		
Cramér test	20.63	0.74
M test	0.85	0.86
T test	0.90	0.93
Wilcox test	0.98	$W > 0.485349$
Quality index test Asym.	0.41	0.17
Quality index test Boots.	0.41	0.21
Zhang test	0.47	0.49

TABLE 4.11 Summary of the univariate and multivariate serial dependence results for Nottawasaga River (station 02ED003).

Univariate Wald–Wolfowitz test				
	$\lvert U \rvert$	**Threshold**	**p-value**	
Q	0.64	1.96	0.52	
V	0.020	1.96	0.98	
Multivariate copula based test				
	Global Cramér-von Mises statistic I_n	**p-value**	***p*-value from Fisher's rule**	***p*-value from Tippet's rule**
(Q, V)	0.0043	0.24	0.81	0.95

(the results are for the year 1999, but we obtained similar results of acceptance for all the appropriate years where the hetereogeneity is suspected).

As it can also be seen from Table 4.11 and Fig. 4.5, the serial independence can be accepted by all the tests.

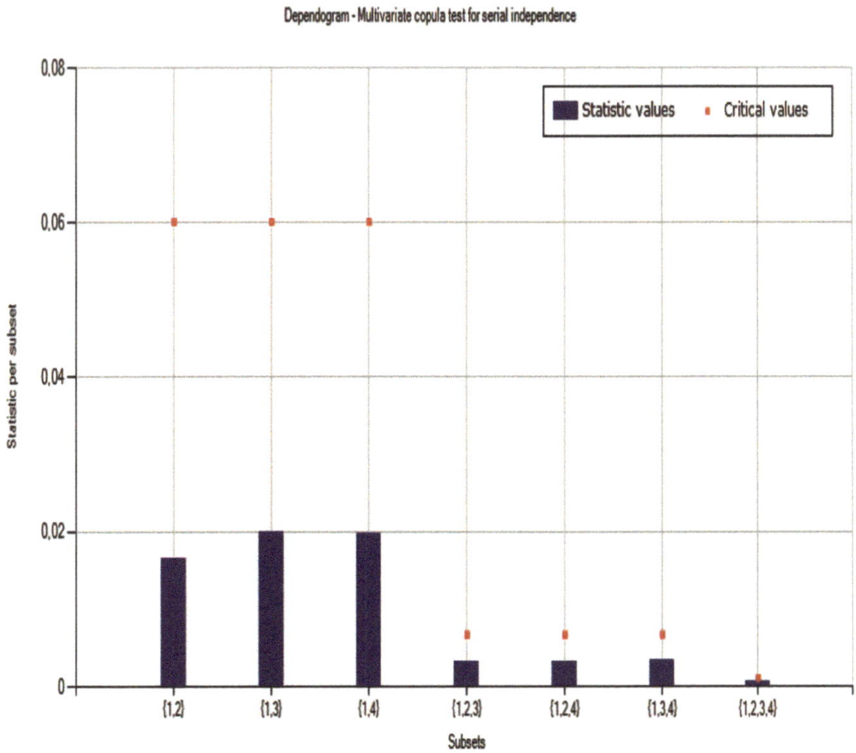

FIGURE 4.5 Dependogram of the subsets for the (Q, V) couple for Nottawasaga River (station 02ED003).

References

Alvo, M., & Park, J. (2002). Multivariate non-parametric tests of trend when the data are incomplete. *Statistics & Probability Letters*, *57*(3), 281–290.

Baringhaus, L., & Franz, C. (2004). On a new multivariate two-sample test. *Journal of Multivariate Analysis*, *88*(1), 190–206.

Beaulieu, C., Ouarda, T. B. M. J., & Seidou, O. (2007). A review of homogenization techniques 559 for climate data and their applicability to precipitation series. *Hydrological Sciences Journal*, *52*(1), 18–37, 560.

Ben Nasr, I., & Chebana, F. (2019). Homogeneity testing of multivariate hydrological records, using multivariate copula L-moments. *Advances in Water Resources*, 134103449.

Beran, R., Bilodeau, M., & de Micheaux, P. L. (2007). Nonparametric tests of independence between random vectors. *Journal of Multivariate Analysis*, *98*(9), 1805–1824.

Blum, J. R., Kiefer, J., & Rosenblatt, M. (1961). Distribution free tests of independence based on the sample distribution function. *Sandia Corporation*.

Bücher, A., Fermanian, J. D., & Kojadinovic, I. (2019). Combining cumulative sum change-point detection tests for assessing the stationarity of univariate time series. *Journal of Time Series Analysis*, *40*(1), 124–150.

Chebana, F. (2013). Multivariate analysis of hydrological variables. *Encyclopedia of Environmetrics*. Available from https://doi.org/10.1002/978047005733.

Chebana, F., & Ouarda, T. B. (2021). Multivariate non-stationary hydrological frequency analysis. *Journal of Hydrology, 593*125907.

Chebana, F., Aissia, M.-A. B., & Ouarda, T. B. (2017). Multivariate shift testing for hydrological variables, review, comparison and application. *Journal of Hydrology, 548*, 88–103.

Chebana, F., Ouarda, T. B., & Duong, T. C. (2013). Testing for multivariate trends in hydrologic frequency analysis. *Journal of Hydrology, 486*, 519–530.

Clement, L., & Thas, O. (2009). Nonparametric trend detection in river monitoring network data: A spatio-temporal approach. *Environmetrics (London, Ont.), 20*(3), 283–297.

El-Shaarawi, A. H. (1993). Environmental monitoring, assessment and prediction of change. *Environmetrics (London, Ont.), 4*(4), 381–398.

Fernández, V. A., Gamero, M. J., & Garcia, J. M. (2008). A test for the two-sample problem based on empirical characteristic functions. *Computational Statistics & Data Analysis, 52* (7), 3730–3748.

Genest, C., & Rémillard, B. (2004). Tests of independence and randomness based on the empirical copula process. *Test, 13*, 335–369.

Ghoudi, K., Kulperger, R. J., & Rémillard, B. (2001). A nonparametric test of serial independence for time series and residuals. *Journal of Multivariate Analysis, 79*(2), 191–218.

Gu, H., Yu, Z., Li, G., & Ju, Q. (2018). Nonstationary multivariate hydrological frequency analysis in the upper Zhanghe River Basin, China. *Water, 10*(6), 772.

Herwartz., & Maxand. (2018).) Nonparametric tests for independence: A review and comparative simulation study with an application to malnutrition data in India. *Statistical Papers.*

Hofert, M., Kojadinovic, I., Mächler, M., & Yan, J. (2018). *Elements of copula modeling with R.* Springer International Publishing.

Karahacane, H., Meddi, M., Chebana, F., & Saaed, H. A. (2020). Complete multivariate flood frequency analysis, applied to northern Algeria. *Journal of Flood Risk Management, 13*(4)e12619.

Khaliq, M. N., Ouarda, T. B. M. J., Ondo, J. C., Gachon, P., & Bobée, B. (2006). Frequency analysis of a sequence of dependent and/or non-stationary hydro-meteorological observations: A review. *Journal of Hydrology, 329*(3-4), 534–552.

Kojadinovic, I., & Yan, J. (2011). Tests of multivariate serial independence based on a Möbius decomposition of the independence empirical copula process. *Annals of the Institute of Statistical Mathematics, 63*, 347–373.

Kundzewicz, Z. W., & Robson, A. J. (2004). Change detection in hydrological records—A review of the methodology/revue méthodologique de la détection de changements dans les chroniques hydrologiques. *Hydrological Sciences Journal, 49*(1), 7–19.

Libiseller, C., & Grimvall, A. (2002). Performance of partial Mann-Kendall tests for trend detection in the presence of covariates. *Environmetrics (London, Ont.), 13*(1), 71–84.

Lund, R., & Reeves, J. (2002). Detection of undocumented change points: A revision of the two-phase regression model. *Journal of Climate, 15*(17), 2547–2554.

Mathur, S. K. (2009). A new nonparametric bivariate test for two sample location problem. *Statistical Methods and Applications, 18*(3), 375.

Oja, H., & Randles, R. H. (2004). Multivariate nonparametric tests. *Statistical Science, 19*(4), 598–605.

Quessy, J. F., Saïd, M., & Favre, A. C. (2013). Multivariate Kendall's tau for change-point detection in copulas. *Canadian Journal of Statistics, 41*(1), 65–82.

Rao, A. R., Hamed, K. H., & Chen, H.-L. (2003). *Nonstationarities in hydrologic and environmental time series.* Dordrecht: Kluwer Academic Publishers.

Salas, J.D. (1993). Analysis and modeling of hydrologic time series. D.R., Maidment, ed. In: The McGraw Hill Handbook of Hydrology (72 p.), Chapter 19.

Santhosh, D., & Srinivas, V. V. (2013). Bivariate frequency analysis of floods using a diffusion based kernel density estimator. *Water Resources Research, 49*(12), 8328–8343.

Serinaldi, F., Kilsby, C. G., & Lombardo, F. (2018). Untenable nonstationarity: An assessment of the fitness for purpose of trend tests in hydrology. *Advances in Water Resources, 111*, 132–155.
Sonali, P., & Kumar, D. N. (2013). Review of trend detection methods and their application to detect temperature changes in India. *Journal of Hydrology, 476*, 212–227.
Wong, H., Hu, B. Q., Ip, W. C., & Xia, J. (2006). Change-point analysis of hydrological time series using grey relational method. *Journal of Hydrology, 324*(1-4), 323–338.
Xiong, L., Jiang, C., Xu, C. Y., Yu, K. X., & Guo, S. (2015). A framework of change-point detection for multivariate hydrological series. *Water Resources Research, 51*(10), 8198–8217.

CHAPTER

5

Modeling in multivariate hydrological frequency analysis with copula

5.1 Introduction

As discussed in Chapter 2, where more details are given, hydrological events are naturally multivariate (e.g., volume and peak for floods) and dependence may be significant between their characteristics/features. Therefore, adopting the multivariate hydrological frequency analysis (HFA) framework to treat those extreme events, as opposed to univariate framework, was supported by several studies especially for a more accurate risk assessment (e.g., Karahacane et al., 2020; Zhu et al., 2019 and references therein). Indeed, univariate HFA is able to provide only a limited assessment of extreme events, ignores the dependence of the characteristics of each event, and reduces the estimation accuracy of the associated risk. The joint study of the characteristics of the events leads to an improved understanding of the hydrological phenomenon. Hence, multivariate studies help to improve the accuracy of estimates and provide information on the pattern of dependence between the characteristics of hydrological events (e.g., Genest & Chebana, 2017; Hao & Singh, 2016). Note that in a multivariate context, there is a cross-dependence between the variables as well as a serial-dependence between the observations. In this chapter, we deal with cross-dependence whereas the serial-dependence was mainly discussed in Chapter 4. In addition, modern multivariate modeling is based on the notion of copula, which is a key tool for modeling the dependence between hydrologic variables in the context of multivariate HFA. Table 5.1 summarizes the different frameworks from univariate to multivariate. In the latter, we have on the one hand distribution-based framework, which includes in particular

Multivariate Frequency Analysis of Hydro-Meteorological Variables
DOI: https://doi.org/10.1016/B978-0-323-95908-7.00006-2

TABLE 5.1 Different frameworks, their limitations, and advantages.

	Univariate	Multivariate		
		Multivariate distributions		
		Normal	Non-normal	Copulas
Advantages	• It is appropriate when a given single variable is the objective of the study • It is usefull when the dependence between variables is not significant • The study results can be used as part of the multivariate HFA study (especially based on copulas)	• They improve the accuracy of estimates and provide information on the pattern of dependence between the characteristics of hydrological events • The joint study of the characteristics of the event leads to a better understanding of the phenomenon		• They are an ingredient for generating joint distributions • They offer a great flexibility for modeling multivariate samples (not linked to the margins which are not necessary similar or in the same family) • The variables do no need to be transformed to fit into a classical framework • This is convenient for interpretation and practical purposes • Copula-based approach separates a multivariate distribution into margins and a copula • This framework takes advantage of univariate HFA results • They capture dependence more broadly • They are useful for applications in simulation and Monte Carlo studies • They improve the accuracy of estimates and provide information on the pattern of dependence between the characteristics of hydrological events • The joint study of the characteristics of the event leads to a better understanding of the phenomenon
Limitations	• It ignores the dependence of the event characteristics • It reduces the accuracy of the estimation of the risk event	• Usually, they are not adapted to study extreme events	• They are much less flexible since they often require the margins to be of the same family	• Mikosch (2006) provided more details, arguments and discussions

(Continued)

TABLE 5.1 (Continued)

	Univariate	Multivariate		
		Multivariate distributions		Copulas
		Normal	Non-normal	
		• They cover only limited shapes of the dependence structure	• Only few distributions have their multivariate extensions available	

normal and non-normal distributions, and on the other hand, the copula-based framework.

Copulas can be used as an essential ingredient in a "recipe" for constructing joint multivariate distributions by combining available marginal distributions according to a specified form of a copula function. Hence, they offer a great flexibility for modeling multivariate samples in a variety of situations. In simple words, a copula is a multivariate distribution function where the margins are uniform over the unit interval [0, 1]. On the other hand, recall that a continuous random variable can be transformed to a uniform random variable over [0, 1] using its probability integral transformation. Consequently, one of the aims of copula is to construct new multivariate distributions by linking different margins. In this way, a multivariate distribution is decomposed into two main components: a copula and the margins. This decomposition provides a very flexible framework in multivariate modeling. Besides constructing multivariate distributions, copulas aim to describe the dependence structure in multivariate data sets.

The main advantage of copulas over classical multivariate distributions is that the dependence between variables can be modeled separately from their marginal distributions, and thus allow to consider margins of different classes. Indeed, given a univariate distribution (e.g., Gamma, Gumbel, or Log-Normal) for each variable (margin), they can be "connected" to each other with a suitable copula. This benefit is particularly useful in hydrology and HFA, where variables describing a given event, such as flood peak and volume, are usually different from each other in their nature, scale, and distribution. In addition, with copula models there is no need to transform variables in order to fit into a classical framework. Dealing directly with the original variables is preferable for practical aspects and interpretation. Another interesting attractive feature of copula is the ability to build on the common and available tools of univariate HFA. Copulas are especially appealing because they capture dependence more broadly than the standard multivariate normal framework and the usual dependence measures. Simulation and Monte Carlo studies represent one of the principal applications of copulas where the generated series are useful for statistical studies such as comparison of a proposed statistical method with its competitors, evaluating robustness properties, or checking the agreement between asymptotic and finite sample results. Despite all the previous advantages, there are some rare critics towards copulas, for instance, in the paper with discussion by Mikosch (2006).

Copulas are receiving increasing attention in a large number of application fields. Modeling dependence structures by copulas is a topic of current research and of recent use in several areas, such as financial assessments, insurance, water science, and hydrology. Although copula approach and modeling have been considered and developed in hydrology during the last two decades only, the corresponding literature is already extensive. For more description of the variety of hydrological and water resources studies dealing with copulas, we suggest the recent works by Hao and Singh (2016), Genest and Chebana (2017), and Zhang and Singh (2019). The reader can also refer to Chapter 2. More precisely, copulas have been increasingly used in multivariate HFA for various multivariate hydrologic analyses where they have been used for precipitations, droughts, floods, and storms. In addition, copulas are employed to study a single feature but at various sites, such as flood peak from two confluent river branches.

Given their importance, advantages, and the increase of the applications, a number of entire books, chapters, and review papers are available covering different points of view related to copulas from statistical theory, computational, variety of application fields including in particular hydrology and HFA. We refer to some of them as a complement for different directions. In terms of mathematical and theoretical developments and the foundations of copula, we refer to Joe (2014) textbook. The latter summarizes the existing books on copulas prior to 2014, including for instance Salvadori et al. (2007). When it comes to computational as well as practical aspects with an interesting variety of topics and discussions, the book by Hofert et al. (2018) is an excellent resource. Chebana (2013), as a chapter, dealt with multivariate HFA including copula as a part of it. However, Hao and Singh (2016) can be seen as an interesting review focusing on copula in water resources and including HFA but not limited to. More recently, the chapter by Genest and Chebana (2017) is dedicated to copula with a focus on HFA. The most recent books are those by Zhang and Singh (2019) and Chen and Guo (2019), which are more oriented to hydrological perspectives including different types of applications such as HFA. From other fields, for instance, Trivedi and Zimmer (2007) reviewed the use of copulas in econometrics, finance, and insurance for practitioners.

Copula models are formally introduced in Section 5.2, whereas in Section 5.3, a number of well-known classes of copula are presented. Section 5.4 deals with dependence measures. Sections 5.5 and 5.6 respectively treat the parameter estimation and the copula model selection.

5.2 Description of copula models

For simplicity, in the following we focus on the bivariate case. However, the majority of the materials presented below can be formulated for higher dimensions. If it is not the case, we provide the appropriate information and references.

Let X and Y be two random variables, such as flood peak and volume. We consider samples of the couple (X, Y) with size n denoted by (x_1, y_1), ..., (x_n, y_n). Let F be the joint distribution function of (X, Y) and F_X and F_Y be the marginal distribution functions respectively of X and Y. Following the Sklar's theorem (Sklar, 1959), F can be decomposed into a marginal distribution corresponding to each variable as well as copula to describe the dependence structure between the involved variables. In other words, Sklar's theorem, as a key result, provides the relationship between a multivariate distribution and the corresponding copula with the marginal distributions. More precisely, Sklar's result states that there exists a copula C such that F can be expressed as:

$$F(x, y) = C\big(F_X(x),\ F_Y(y)\big) \text{ for all real } x \text{ and y} \tag{5.1}$$

In the case where the margins F_X and F_Y are continuous, which is common in hydrology, the copula C is unique.

The copula C is basically a function that allows to connect the multivariate joint distribution to its margins. It captures the essential of the dependence structure between variables, which is more informative than the dependence measures (e.g., Kendall's tau and Spearman's rho, defined below).

Formally, a function $C: I \times I \rightarrow I$ $(I = [0, 1])$ is said to be a copula if the following conditions are satisfied:

- for all $u, v \in I$: $C(u, 0) = C(0, \mathrm{v}) = 0$, $C(u, 1) = u$, and $C(1, v) = v$;
- for all $u_1,\ u_2,\ v_1,\ v_2\ \in I$
  $u_1 \leq u_2$ and $v_1 \leq v_2$: $C(u_2, v_2) - C(u_2, v_1) - C(u_1, v_2) + C(u_1, v_1) \geq 0$

The first condition indicates that the margins are uniform, whereas the second condition is to ensure that the copula is 2-increasing. Both conditions are adapted formulations of conditions of a classical multivariate function to be a cumulative distribution function.

After Sklar's theorem, one of the important theoretical results is the Fréchet–Hoeffding bounds which states that:

$$\max(u + v - 1, 0) \leq C(u, v) \leq \min(u, v) \text{ for } 0 < u, v < 1 \tag{5.2}$$

It is illustrated in Fig. 5.1 for a given copula.

A copula C is called *absolutely continuous* if its density function c exists and is integrable and it is given by

$$c(u, v) = \frac{\partial^2}{\partial u \partial v} C(u, v) \text{ for } 0 < u, v < 1$$

For formulations of the above expressions in the d-dimension, see for example, Hofert et al. (2018) or Zhang and Singh (2019). Fig. 5.2 illustrates density functions of some well-known copulas with different shapes but

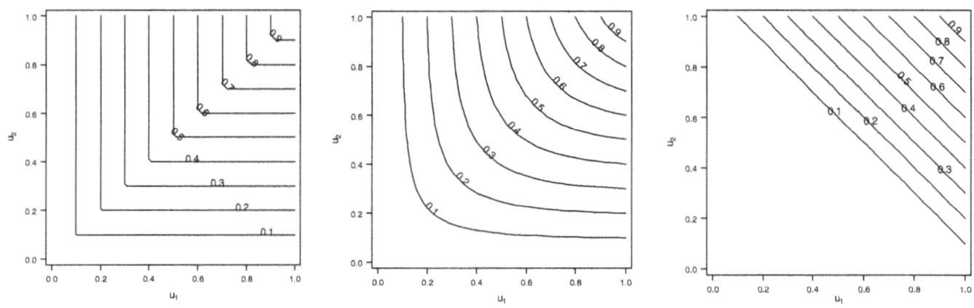

FIGURE 5.1 Illustration of Fréchet–Hoeffding bounds (left and right) with Frank copula and parameter = 4 (middle).

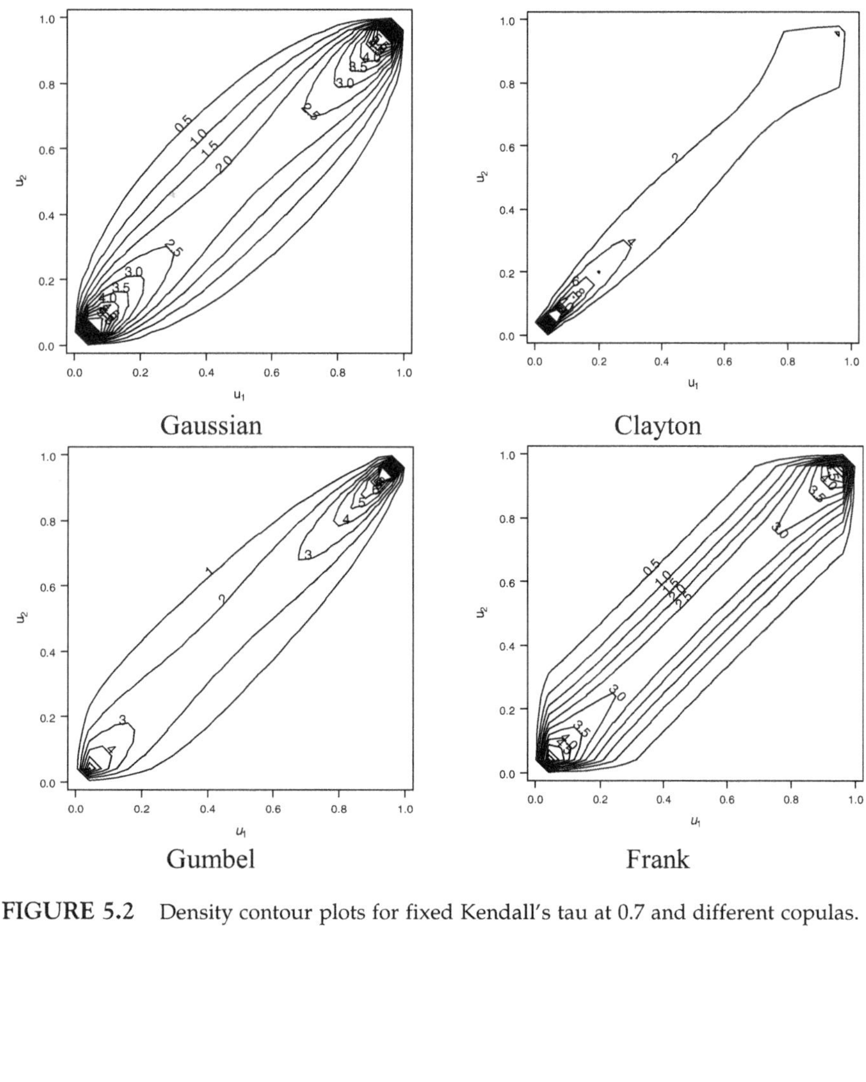

FIGURE 5.2 Density contour plots for fixed Kendall's tau at 0.7 and different copulas.

for the same value of Kendall's tau. This shows that the information provided only by Kendall's tau is not enough to represent the dependence structure which is not the case when using copula.

The copula C as a model has its empirical version called empirical copula defined by

$$\hat{C}_n(u,v) = \frac{1}{n}\sum_{i=1}^{n} 1\left(\hat{U}_i \le u,\ \hat{V}_i \le v\right) \text{ for } 0 < u, v < 1 \tag{5.3}$$

which is based on the so-called pseudo-observations given by

$$\left(\hat{U}_i, \hat{V}_i\right) = (F_{nX}(X_i), F_{nY}(Y_i)),\ \ i = 1, \ldots, n \tag{5.4}$$

where F_{nX} is the empirical distribution function of the margin X (similarly for Y) given by

$$F_{nX}(x) = \frac{1}{n+1}\sum_{i=1}^{n} 1(X_i \le x) \text{ for all real x} \tag{5.5}$$

where $1(A)$ is the indicator of the set A, that is, $1(A) = 1$ when the event A occurs and $1(A) = 0$ otherwise. Note that $(n + 1) \times F_{nX}(X_i)$ is the rank of X_i among $X_1, \ldots, X_n$ (respectively for Y) and all the information needed to identify C is contained in the joint pattern described by $(\hat{U}_i, V_i)$, $i = 1, \ldots, n$. The empirical copula(5.3) is a consistent estimator of C whose asymptotic results follow from those of the *empirical copula process* $\sqrt{n}(\hat{C}_n(u,v) - C(u,v))$ $u, v \in (0, 1)$. More details can also be found, for instance, in Genest and Chebana (2017) and Hofert et al. (2018) including formulations for d-dimensions.

Constructing copulas is a fundamental topic in copula development. To this end, a number of methods can be considered, such as the inversion method, the geometric method, and the algebraic method. Nelsen (2006, Chapter 3) provided an interesting discussion on these methods with important and useful details for a large number of copula families. More recently, Joe (2014) also dedicated an entire chapter for this topic.

5.3 Classes of copula

A number of copula families are developed and studied in the literature where each of them represents a different dependence structure. Archimedean and extreme value (EV) copulas are among the very popular classes in statistics, in hydrology, as well as in many other application fields. Presented in different ways and different levels of details, lists of copula formulations can be found for instance in Joe (2014),

Salvadori et al. (2007), or Zhang and Singh (2019). In Joe (2014, Chapter 4), one can find a large number of copulas with their tail dependence measures, an interesting list of detailed properties, extensions, as well as historical notes. An interesting short discussion (without formulas) can also be found in Trivedi and Zimmer (2007).

5.3.1 Archimedean copulas

Archimedean copulas are among the well-known and widely considered copulas in a variety of applications including hydrology and water resources. Indeed, they have some attractive properties such as: (1) they can
be easily constructed; (2) an important number of copula families are Archimedean ones; and (3) their mathematical formulations are simple and elegant. For instance, in Nelsen (2006, Chapter 4), a large number (22 copulas) of Archimedean copulas are presented with specific references and notes for each one.

A bivariate Archimedean copula is characterized by the expression:

$$C(u,v) = \varphi^{-1}(\varphi(u) + \varphi(v)), 0 < u, v < 1 \tag{5.6}$$

where $\varphi(.)$, called the generator, is a convex decreasing function with $\varphi(1) = 0$.

Archimedean copulas with one parameter include well-known copulas such as Frank, Clayton, Ali–Mikhail–Haq (AMH), Joe, and Gumbel–Hougaard (also called Gumbel). Table 5.2 provides formulations of the generator of these copulas. For illustrative purposes, for a given value of Kendall's tau, here 0.4, Fig. 5.3 shows the curves of the generators of the copulas in Table 5.2. Note that for this value of tau, the AMH copula is not defined.

TABLE 5.2 Some Archimedean copulas.

Archimedean copulas	$C(u,v)$	$\varphi_\theta(t)$
Frank	$-\frac{1}{\theta}\log\left[1 + \frac{(e^{-\theta u}-1)(e^{-\theta v}-1)}{e^{-\theta}-1}\right], \theta \neq 0$	$-\log\frac{e^{-\theta t}-1}{e^{-\theta}-1}$
Clayton	$\frac{[1-\theta(1-u)(1-v)](1-\theta)+2\theta uv}{[1-\theta(1-u)(1-v)]^3}, -1 \leq \theta \neq 0$	$\frac{1}{\theta}(t^{-\theta}-1)$
Ali–Mikhail–Haq (AMH)	$\frac{uv}{1-\theta(1-u)(1-v)}, -1 \leq \theta < 1$	$\log\frac{1-\theta(1-t)}{t}$
Joe	$1-\left((1-u)^\theta+(1-v)^\theta-(1-u)^\theta(1-v)^\theta\right)^{1/\theta}, \theta \geq 1$	$-\log(1-(1-t)^\theta)$
Gumbel–Hougaard	$\exp\left\{-[(-\log u)^\theta+(-\log v)^\theta]^{\frac{1}{\theta}}\right\}, \theta \geq 1$	$(-\log t)^\theta$

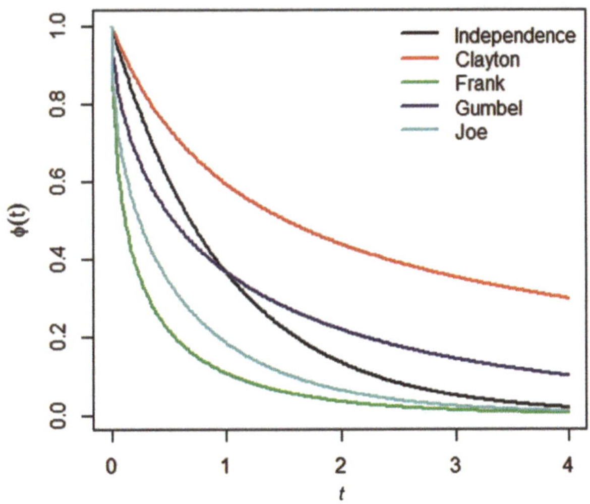

FIGURE 5.3 Generator function for some Archimedean copulas with tau = 0.4 (AMH copula does not exist for this value of tau).

An Archimedean copula has some interesting analytical properties, including:

1. C is symmetric, that is, $C(u,v) = C(v,u)$ for all u and $v \in I$
2. C is associative, that is, $C(C(u,v),w) = C(u, C(\mathrm{v},\mathrm{w}))$ for all u, v, $w \in I$
3. If φ generates C, then $\varphi' = a\varphi$ generates C as well, for a positive constant a.

In other words, Archimedean copulas are symmetric in their arguments and Archimedean definition is preserved with the same generator for all their lower dimensional margins. This feature is useful since adding or deleting a variable does have an impact on the joint dependence structure. However, in some situations, this feature can also be seen as a limitation and as a lack of flexibility. To overcome the latter, one can consider *asymmetric* extensions such as hierarchical Archimedean copulas or Liouville copulas.

For more flexibility, a number of multi-parameter Archimedean copulas are also available in the literature where the reader is referred to Joe (2014) in the general statistical context and in particular to Ben Nasr and Chebana (2019) in the multivariate HFA context. For instance, Archimedean copulas with two-parameters include BB1 (with particular cases as Clayton and Gumbel), BB6 (Joe and Gumbel), BB7 (Joe and Clayton), and BB8 (Joe and Frank).

According to the shape and strength of the dependence, some copulas are more appropriate than others. For instance, Clayton copula is an appropriate choice when correlation between two variables is strong in the left tail whereas it is Gumbel copula when it comes to the right tail. Both Clayton and Gumbel copulas do not allow negative dependence. However, the Frank copula is popular since it covers negative dependence, the dependence is symmetric in both tails, and it is "comprehensive" as its dependence reaches both Fréchet-Hoeffding bounds (5.2). On the other hand, the Frank copula is most appropriate for data that exhibit weak tail dependence.

The Archimedean copula family is widely used in hydrology and in multivariate HFA. Indeed, they were among the first ones used in early 2000 (e.g., Salvadori & De Michele, 2004), later (e.g., Chebana & Ouarda, 2011), and very recently (e.g., Liu et al., 2020). More specifically, from the hydrological literature, Frank copula is largely considered or selected. Frank is followed by Gumbel copula in a number of studies. Other Archimedean copulas are also selected such as Joe and Clayton.

One of the attractive features of Archimedean copulas is their simplicity of formulation in high dimensions. For instance, Grimaldi and Serinaldi (2006) have been interested in the three-variate Archimedean copulas to study rainfall, drought, and flood events whereas the fourth dimension was studied in Ayantobo et al. (2019). In the case where the symmetry of Archimedean copulas is restrictive, their asymmetric extensions have been also introduced and studied in hydrology (e.g., Ma et al., 2013; Serinaldi & Grimaldi, 2007).

The Kendall's function K defined as $K(t) = P\,[C(U,V) \leq t]$, $t \in [0,1]$ is particularly useful for Archimedean copulas. However, this function is *not* appropriate for EV copulas (defined below), as all EV copulas have the same Kendall's function. An Archimedean copula, with a generator function φ, is characterized by the following Kendall's function:

$$K_\varphi(t) = t - \frac{\varphi(t)}{\varphi'(t)} \tag{5.7}$$

which can be estimated by:

$$\hat{K}_n(t) = \frac{1}{n}\sum_{i=1}^{n} 1_{[w_i \leq t]} \text{ where } w_i = \frac{1}{n-1}\sum_{t=1}^{n} 1_{[x_t < x_i, y_t < y_i]},\ i = 1, .., n \tag{5.8}$$

for a given bivariate sample $(x_1, y_1), (x_2, y_2), \ldots, (x_n, y_n)$. The functions K_φ and $\hat{K}_n$ can be used to construct goodness-of-fit tests for Archimedean copulas (see Section 5.6). Fig. 5.4 shows the Kendall's function for some Archimedean copulas with different values of Kendall's tau. We observe that the higher the tau, the close to the diagonal is the function K (corresponding to perfect dependence).

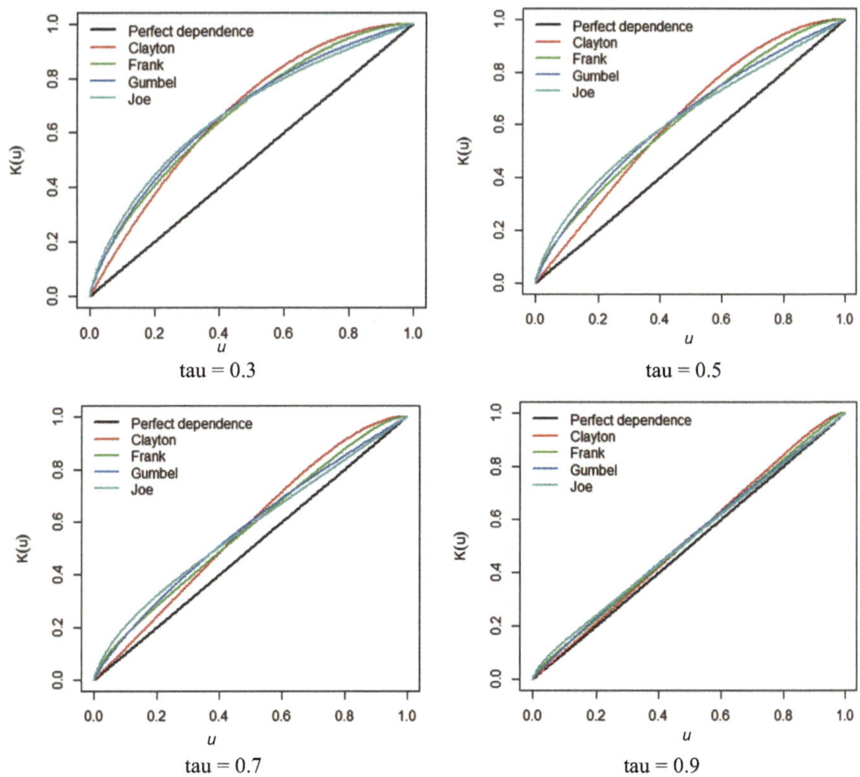

FIGURE 5.4 The function K for some Archimedean copulas with different values of tau τ.

5.3.2 Extreme-value copulas

Extreme value (EV) copulas is another class of interest in HFA among other applications dealing with risk assessment. The definition of EV copula is based a dependence function A as:

$$C(u,v) = \exp\left\{ (\log u + \log v) A\left(\frac{\log u}{\log u + \log v} \right) \right\}, 0 < u, v < 1 \qquad (5.9)$$

where the Pickands dependence function A is convex and defined on $[0, 1]$ with $\max\{t, 1-t\} \le A(t) \le 1$. Note that the function A plays a similar role for EV copulas as the generator $\varphi(.)$ for Archimedean copulas.

Equivalently, a copula C is said to be EV if,

$$C^t\left(u^{1/t}, v^{1/t}\right) = C(u,v) \text{ for all } \quad t > 0 \text{ and } 0 < u, v < 1 \qquad (5.10)$$

This property is particularly suitable for block maxima data series such as in HFA with yearly maxima of daily flow measurements.

EV copulas include for instance Galambos, Hüsler–Resiss, Gumbel, and Tawn (see Table 5.3). The Gumbel copula, at the same time Archimedean and EV, is a typical example in hydrology. It is widely used to model flood volume and peak dependence. Although EV copulas, from their definition, can be asymmetric, the EV dependence structure is inherited by all subsets of a vector whose copula satisfies property (5.10). See Genest and Chebana (2017) for expressions with d-dimensions and Genest and Nešlehová (2012b) for a review and more details regarding this class of copulas.

In Fig. 5.5, we present the function A for two EV copulas (i.e., Galambos and Gumbel) for different values of Kendall's tau. We observe that there are almost no differences between the functions A of these copulas given the same value of Kendall's tau, as also observed in Hofert et al. (2018) for other copulas.

TABLE 5.3 Some extreme value copulas and their Pickands functions.

Copula family	Pickands function $A(t)$, t in (0, 1)	Parameter range
Galambos	$1-\left(t^{-\theta}+(1-t)^{-\theta}\right)^{-1/\theta}$	$\theta\in(0,\infty)$
Hüsler–Resiss	$t\Phi\left(\theta+\frac{1}{2\theta}\log\left(\frac{t}{1-t}\right)\right)+(1-t)\Phi\left(\theta+\frac{1}{2\theta}\log\left(\frac{1-t}{t}\right)\right)$	$\theta\in(0,\infty)$
Gumbel–Hougaard	$\left(t^{\theta}+(1-t)^{\theta}\right)^{1/\theta}$	$\theta\in[1,\infty)$
Tawn	$\theta t^2-\theta t+1$	$\theta\in[0,1]$

Φ is the cumulative distribution function of the standard normal distribution.

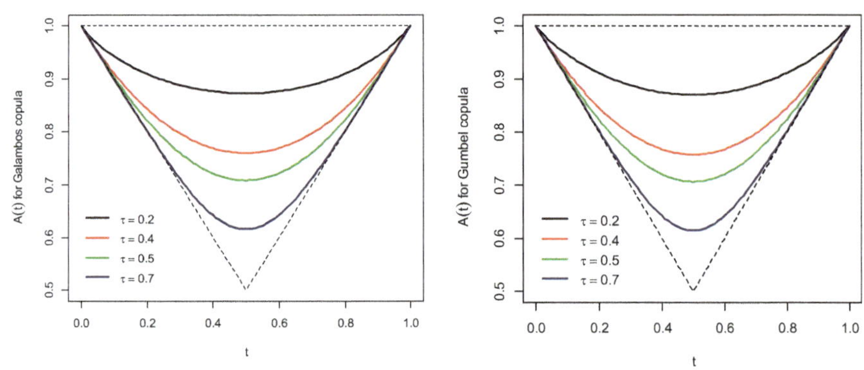

FIGURE 5.5 The function A for Galambos (left) and Gumbel (right) copulas with different values of Kendall's tau.

In the bivariate case, the dependence function A can be estimated (nonparametrically) by

$$\hat{A}_n^P(t) = \left\{ \frac{1}{n} \sum_{i=1}^{n} \hat{\xi}_i(1-t) \right\}^{-1}, \quad 0 < t < 1 \tag{5.11}$$

where $\hat{\xi}_i(t) = \min\{-\ln(\hat{U}_i)/t, -\ln(\hat{V}_i)/(1-t)\}$, $i = 1, \ldots, n$. This estimator is a rank-based version of the original Pickands estimator.

An alternative estimator of A is the Capéraà, Fougères and Genest (CFG) estimator with a formulation in the bivariate case given by

$$\hat{A}_n^{CFG}(t) = \exp\left[-\gamma - \frac{1}{n} \sum_{i=1}^{n} \log \hat{\xi}_i(1-t) \right], \tag{5.12}$$

where $\gamma \approx 0.577$ is the Euler constant. Under some regularity conditions, both estimators in (5.11) and (5.12) are consistent and are asymptotically normal at every fixed $t \in (0, 1)$. However, for finite sample sizes from simulations, the CFG estimator generally outperforms the Pickands estimator.

EV copulas are crucial in hydrological studies especially in HFA since the main aim of HFA is to estimate hydrological risks, which occur usually as extreme events. EV copulas have been used or selected to study multivariate hydrological events like floods and droughts (e.g., Ben Ben Alaya et al., 2018; Papaioannou et al., 2016). The Gumbel copula is one of the most selected copulas since it shares interesting features of both classes (EV and Archimedean). More specifically, Gumbel copula is considered or selected, for instance, in Salvadori and De Michele (2010) and Sharma and Mujumdar (2019). The reader is referred to Salvadori & De Michele (2011) for EV copula applications in hydrology.

5.3.3 Meta-elliptical copulas

Meta-elliptical copulas are derived from elliptical distributions. These copulas have a good compromise between convenience and flexibility. A random vector X is said to have an elliptical *distribution* with mean μ and dispersion matrix $\Sigma = AA^T$, where A^T denotes matrix transposition, if X can be written as $X = \mu + RAZ$, where R is a strictly positive continuous random variable, A is a 2×2 matrix of constants, and Z is a random vector independent of R and has a uniform distribution on the unit sphere $T_2 = \{(t_1, t_2): t_1^2 + t_2^2 = 1\}$. The density function of an elliptical distribution (if it exists) is given by

TABLE 5.4 Some meta-elliptical copulas and their generator functions.

Copula	Generator function $g(t)$
Normal	$(2\pi)^{-1}\exp\left(-\frac{t}{2}\right)$
Student	$\frac{\Gamma\left(\frac{d+v}{2}\right)}{\Gamma\left(\frac{d}{2}\right)\pi v}\left(1+\frac{t}{v}\right)^{-\frac{2+v}{2}}$
Cauchy	$\frac{\Gamma\left(\frac{3}{2}\right)}{\pi\Gamma\left(\frac{1}{2}\right)}(1+t)^{-\frac{3}{2}}$

$$f(x) = |\Sigma|^{-1/2} g\left((x-\mu)^T \Sigma^{-1}(x-\mu)\right) \tag{5.13}$$

where the function g: $R \rightarrow R^+$ is called a density generator.

The associated copula to the distribution of X is called meta-elliptical copula. This copula can be obtained with the inversion method. Indeed, if F is a joint bivariate elliptical distribution, with univariate marginal distribution functions F_X and F_Y, then the corresponding elliptical copula is given by

$$C(u, v) = F\left({F_X}^{-1}(u), {F_Y}^{-1}(v)\right)$$

In statistics, a number of studies focused on meta-elliptical copula properties. The normal copula is the first and direct example of this class. It corresponds to the case where R^2 is chi-square distributed with two degrees of freedom. The multivariate Student t copula as well as the Cauchy copula also belong to this class (see Table 5.4). Other examples, based on the generator function g, can be found in Zhang and Singh (2019, Table 7.1). See Genest and Chebana (2017) for more details as well as for the d-dimension formulations.

Because of their flexibility and relation to normal distribution, meta-elliptical copulas have been employed in hydrology such as in droughts and extreme rainfall (Ma et al., 2013; Song & Singh, 2010). They have been considered but not necessary selected in some other studies.

5.3.4 Other classes of copulas

The above classes of copulas are the most developed and used in statistics as well as in hydrology among other fields of applications. However, extensions of some of the previous copulas developed such as asymmetric copulas, hierarchical Archimedean copulas, or Liouville copulas, are also available in the literature and considered in hydrological applications (e.g., Ayantobo et al., 2019; Grimaldi & Serinaldi, 2006). On the other hand, a number of other classes are also available such as the Farlie–Gumbel–Morgenstern (FGM), Plackett, Entropy, and vine copulas.

The FGM copula can be seen as a perturbation of the independence copula as

$$C(u,v) = uv[1 + \theta(1-u)(1-v)],\ -1 \le \theta \le 1 \tag{5.14}$$

Its simplicity makes it attractive whereas it is restrictive because it is only useful for modest dependence between the margins. It is considered in a limited number of hydrological studies (e.g., Papaioannou et al., 2016).

The Plackett copula is defined as

$$C\left(u,v\right) = \left(1 + (\theta-1)(u+v) - \sqrt{[1+(\theta-1)(u+v)]^2 - 4\theta(\theta-1)uv}\right)/(2(\theta-1)),\ \ \theta \ge 0 \tag{5.15}$$

Probably not as much as Archimedean and EV copulas, Plackett copulas have been used in hydrology in a number of studies such as Kao and Govindaraju (2007) and Papaioannou et al. (2016).

The entropy copula combines the principle of maximum entropy and copula. It is useful when no prior information about the distribution function is available. A few hydrological applications considered entropy copulas, such as Piantadosi et al. (2012), and Li and Zheng (2016).

Vine copulas are very flexible, especially for high dimensions, which are built from a tree-like structure of bivariate copulas. They consider conditional distributions in more than one level of the hierarchy. An attractive property of vine copulas is that different subgroups of variables do not necessary have the same dependence structure. However, the cost of this flexibility is more complex in inference procedures. In hydrology, this class of copulas is starting to attract attention especially because of the need to treat some events in high dimensions where the other classes were restrictive. Recent studies can be found such as Shafaei et al. (2017) and Tosunoglu and Singh (2018) where the focus is not always in frequency analysis but in a variety of hydrological problematics. Given the growing attention to vine copulas, it is important to note the entire recent book dedicated to their study by Czado (2019).

5.4 Dependence measures

When we talk about (cross) dependence, the first thing to come in mind is to evaluate some dependence coefficients. They summarize the dependence, although limited compared to copulas, and are useful. Here, we present the most used ones, that is, Pearson's rho ρ_P, Kendall's tau τ_K, Spearman's rho ρ_S, and upper-lower tail dependence coefficients. However, other dependence measures are available in the literature including Blomqvist's beta, Gini's gamma, and Blest's coefficient.

5.4.1 Overall dependence measures

Usually, the dependence can be measured by a number of coefficients, in particular Pearson's correlation rho ρ_P, Kendall's tau τ_K, and Spearman's rho ρ_S. These dependence measures are the most known and employed in statistics as well as in applications. An interesting discussion and results on the relationship between Kendall's tau and Spearman's rho can be found in Nelsen (2006, Chapter 5).

The empirical version of the Pearson correlation coefficient ρ_P is given by (among other equivalent expressions):

$$\rho_{P,n} = \frac{n\sum_{i=1}^{n} X_i Y_i - \sum_{i=1}^{n} X_i \sum_{i=1}^{n} Y_i}{\sqrt{n\sum_{i=1}^{n} X_i^2 - \left(\sum_{i=1}^{n} X_i\right)^2}\sqrt{n\sum_{i=1}^{n} Y_i^2 - \left(\sum_{i=1}^{n} Y_i\right)^2}} \tag{5.16}$$

Even though it is widely used in practice, the Pearson correlation coefficient has a number of limitations such as it is not margin-free, it is closely related to Gaussian distribution, and linearity.

The definition of Kendall's tau τ_K from a sample (empirical version) is given by

$$\tau_n = 1 + \frac{4}{n(n-1)}\sum_{i=1}^{n}\sum_{j=1}^{i-1} 1\{(X_i - X_j)(Y_i - Y_j) > 0\} \tag{5.17}$$

Note that an equivalent formulation of this empirical version of Kendall's tau can also be obtained on the basis of the observation ranks.

The empirical formula of the Spearman's ρ_S is:

$$\rho_{S,n} = 1 - \frac{6}{n(n^2-1)}\sum_{i=1}^{n}(R_i - S_i)^2 \tag{5.18}$$

where $R_1, \ldots, R_n$ and $S_1, \ldots, S_n$ represent the ascending ranks associated with the samples of X and Y, respectively. Other formulations can also be find in a number of copula textbooks, for example, Hofert et al. (2018). In the bivariate case, Spearman's rho ρ_S can be seen as a Pearson's correlation coefficient based on the ranks of the observations.

Since a copula function is the most complete way to describe the dependence structure, it is useful to find the relationship between the copula parameter, say θ, and each one of the coefficients τ_K and ρ_S (which is not the case for ρ_P since it is not margin-free). The population versions of the last two coefficients can also be obtained respectively in terms of copula C as:

$$\tau_K = 4\int_{[0,1]^2} C(u,v)\mathrm{d}C(u,v) - 1 \tag{5.19}$$

$$\rho_S = 12\int_{[0,1]^2} uv\,\mathrm{d}C(u,v) - 3 \tag{5.20}$$

where these coefficients can be seen as moments of the copula. Table 5.5 gathers the formulations of these coefficients for a number of copulas, well-known and used in hydrology. These relations are established, explicitly or through numerical approximations. In Zhang and Singh (2019), one can find a list of a large number of Archimedean copulas and their relations with Kendall's tau. In addition, d-dimension definitions are also available for these coefficients (see, e.g., Zhang & Singh, 2019, Table 4.6). These relations are useful to the estimation of the copula parameter (see Section 5.5, moment method).

One of the advantages of considering these coefficients is that they are prior to selecting the copula. Hence, this is useful for preliminary copula selection where some copula families for which the values of τ_K and ρ_S range in restricted intervals which makes the corresponding copulas inadequate in some contexts. As an example, in the FGM bivariate copula, we have $\tau_K \in [-2/9, 2/9]$. Other examples include the AMH and Tawn copulas, where respectively $\tau_K \in [(5 - 8\log 2)/3, 1/3] \sim = [-0.1817, 0.3333]$ and $[0, 0.4184]$. As an example for ρ_S, it should be in $[0, 0.5874]$ for the Tawn copula.

where $\mathrm{dilog}(x) = \int_1^x \frac{\ln t}{1-t}dt$ and $D_k(x) = \frac{k}{x^2}\int_0^x \frac{t^k}{e^t - 1}dt$ is the Debye funtion for any positive interger k

Example 1

This is a simple example for the Clayton copula with different values of its parameter (Table 5.6) and the corresponding τ_K and ρ_S along with their respective estimators based on a generated sample of size $n = 100$. We observe that the values of τ_K (and also ρ_S) increase with the values of the parameter θ. On the other hand, overall the estimated values are very close to their respective true values (except a slight difference for ρ_S in the case $\theta = -0.5$).

Example 2

In this second example, we consider the dataset used in Chapter 2 for flood volume V and flood peak Q. Recall that this dataset is from

TABLE 5.5 Kendall's tau and Spearman's rho (τ_K and ρ_S) for some known copulas.

Copula	Kendall's tau	Spearman's rho
Frank	$1-\frac{4}{\theta}[1-D_1(\theta)]$	$1-\frac{12}{\theta}[D_1(\theta)-D_2(\theta)]$
Clayton	$\frac{\theta}{\theta+2}$	Complicated form
Ali–Mikhail–Haq (AMH)	$\frac{3\theta-2}{3\theta}-\frac{2}{3}(1-\frac{1}{\theta})^2\log(1-\theta)$	$\frac{12(1-\theta)}{\theta^2}\mathrm{dilog}(1-\theta)-\frac{24(1-\theta)}{\theta^2}\log(1-\theta)-\frac{3(\theta+12)}{\theta}$
Gumbel–Hougaard	$\frac{\theta-1}{\theta}$	No closed form
Farlie–Gumbel–Morgenstern (FGM)	$\frac{2\theta}{9}$	$\frac{\theta}{3}$
Plackett	Num. Approx.	$\frac{\theta+1}{\theta-1}-\frac{2\theta\log(\theta)}{(\theta-1)^2}$
Galambos	$\frac{\theta+1}{\theta}\int_0^1\left\{t^{-1/\theta}+(1-t)^{-1/\theta}-1\right\}dt$	Num. Approx.
Hüsler–Reiss	Num. Approx.	Num. Approx.
Normal	$\frac{2}{\pi}\arcsin(\theta)$	$\frac{6}{\pi}\arcsin\left(\frac{\theta}{2}\right)$

TABLE 5.6 Values of the parameter θ for Clayton copula and the corresponding values of tau and rho as well as their respective estimates based on a generated sample from Clayton with $n = 100$.

θ	−0.5	2	4	9
τ_K	−0.3333	0.5000	0.6667	0.8182
Estimated τ_K	−0.3253	0.5184	0.6711	0.8307
ρ_S	−0.4670	0.6829	0.8465	0.9506
Estimated ρ_S	−0.4206	0.6900	0.8409	0.9550

TABLE 5.7 Theta for different copulas for data of flood peak and volume (station 02ED003, Ontario, Canada).

	Coefficients	**Values of θ**					
		Clayton	**Frank**	**Gumbel**	**FGM**	**Plackett**	**AMH**
τ_K	0.3245	0.9610	3.2012	1.4805	τ_K must be in [−2/9, 2/9]	4.4910	0.9862
ρ_S	0.4364	0.8637	2.8986	1.4367	ρ_S must be in [−1/3, 1/3]	4.0291	0.9432
		Galambos	**Tawn**	**Hüsler–Reiss**	**Joe**	**Normal**	
τ_K	0.3245	0.7537	0.8132	1.1769	1.8688	0.4880	
ρ_S	0.4364	0.7035	0.7723	1.1140	Does not exist	0.4530	

Nottawasaga River near Baxter (station 02ED003) in Ontario, Canada. The corresponding empirical values for (Q, V) of τ_K and ρ_S are 0.3245 and 0.4364, respectively. Note that the Pearson coefficient in this case is 0.3812. In Table 5.7, we present the corresponding values of the parameter θ of different copulas (estimators based on τ_K and ρ_S relationships with the parameter). We observe that the parameter values for a given copula are slightly different (but in the same range), either obtained from τ_K or ρ_S. It seems also that the FGM copula is not appropriate for this dataset since the values of τ_K and ρ_S are both outside of their respective ranges for this copula. Note also that for Joe copula, the relationship between the parameter and Spearman's rho is not available (not specifically for this dataset).

5.4.2 Tail dependence measures

The previous dependence measures deal with the overall dependence between variables including the central body as well as the tails of the joint distribution. However, the aim of tail dependence coefficients is to focus on quantifying the tail part of this dependence. Tail dependence is a property of copulas and plays an important role in some application fields in particular in hydrology. Indeed, upper tail dependence (UTD) can be seen as the capacity of a copula to connect extreme values, for example, extreme flood peaks to extreme volumes. UTD coefficient, denoted λ_U^C, can be formulated as (if the limit exists):

$$\lambda_U^C = \lim_{w \to 1^-} \frac{1 - 2w + C(w, w)}{1 - w} \tag{5.21}$$

Similarly, the lower tail dependence coefficient can also be defined as $\lambda_L^C = \lim_{w \to 0^+} C(w, w)/w$. Explicit and direct expressions of λ_U^C and λ_L^C can be obtained with respect to the parameters or generators of some common copulas. For instance, one has $\lambda_L{}^C = 0$ but $\lambda^C{}_U = 2{-}2\,A(0.5)$ fora bivariate EV copula with dependence function A. Some of these expressions are presented in Table 5.8, whereas other examples can be found for example, in Salvadori et al. (2007) and Joe (2014).

TABLE 5.8 Upper tail dependence coefficient for different copulas with respect to their parameters.

Copula	λ_U^C
Independence	0
Farlie–Gumbel–Morgenstern	0
Ali–Mikhael–Haq	0
Clayton	0
Frank	0
Normal	0
Plackett	0
Joe	$2 - 2^{1/\theta}$
Gumbel (Archimedean/extreme value)	$2 - 2^{1/\theta}$
Galambos (extreme value)	$2^{-1/\theta}$
Hüsler–Reiss (extreme value)	$2 - 2\Phi(1/\theta)$, Φ is standard normal distribution function

Considering λ_U^C and λ_L^C can sometimes reveal important differences between copulas with overall similar dependence structures. A typical example is the case of the normal copula (null coefficients) and Student *t* copula (strictly positive coefficients). Hence, joint extreme risks can be underestimated via the normal copula.

The direct empirical version of the UTD in (5.19) is not available because of the limit in the formula. Hence, a limited number of estimators of λ_U^C are proposed in the literature. Among them, the nonparametric estimator $\hat{\lambda}_U^{CFG}$ after Capéraà, Fougères, and Genest is given by

$$\hat{\lambda}_U^{CFG} = 2 - 2\exp\left[\frac{1}{n}\sum_{i=1}^{n}\log\left\{-\frac{\sqrt{\log(R_i/(n+1))\log(S_i/(n+1))}}{\log\max(R_i/(n+1), S_i/(n+1))^2}\right\}\right] \tag{5.22}$$

The estimator has been already used in some hydrological studies (e.g., Requena et al., 2013; Serinaldi, 2008). It does not require any tuning parameter as other estimators. It is simple to evaluate and it also performs well even when the assumption that the empirical copula is approximated by an EV copula is not fulfilled.

Alternative multivariate extreme coefficients are considered in the literature. In a hydrological context, Lekina et al. (2015) considered different tail dependence measures where the authors found that first (as expected) the overall dependence measures are inadequate to quantify the extreme risk and the dependence in the tail, and second (more importantly here), the UTD measure could fail to discriminate between the degrees of relative strength of dependence for asymptotically independent variables. Therefore, for an effective risk assessment, the authors recommended to consider more than one tail dependence measure.

Example 1

Based on the dataset on flood volume and peak from Chapter 2, first we evaluated the Kendall's tau as overall dependence measure $\tau_K = 0.3245$ (obtained in the previous example). Then, we estimated the UTD measure as in (5.22) as an empirical nonparametric measure ($\hat{\lambda}_U^{CFG} = 0.3142$). On the other hand, we evaluated the UTD measure for a number of copulas based on the estimation of their respective parameters and the formulas in Table 5.8. The obtained results are presented in Table 5.9. We observe that some copulas are not appropriate for this dataset since the corresponding tail dependence measure is null whereas the empirical estimator is not ($\hat{\lambda}_U^{CFG} = 0.3142$). This is mainly the case of Archimedean copulas, except Joe copula where λ_U^C is non-null but relatively higher than the empirical one. However, the EV

TABLE 5.9 The empirical tail dependence coefficient, Kendall's tau, and estimated parameter for different copulas (station 02ED003, Ontario, Canada).

Copula	θ_n (inverse tau[a])	λ_U^C
AMH	0.9862	0
Clayton	0.9610	0
Frank	3.2012	0
Normal	0.4880	0
Plackett	4.4910	0
Joe	1.8688	0.5510
Gumbel	1.4805	0.4029
Galambos	0.7537	0.3987
Hüsler–Reiss	1.1769	0.3955
FGM	τ_K must be in $[-2/9, 2/9]$	NA

[a] *See Section 5.5 about copula parameter estimation.*

copulas provided non-null values of λ_U^C which are in a range close to the empirical estimator $\hat{\lambda}_U^{CFG}$.

5.5 Copula parameter estimation

One of the main topics in copula is the inference regarding its parameter $\boldsymbol{\theta}$. Several methods of copula parameter estimation have been developed, including the inference function of margins (IFM), the maximum pseudo-likelihood (MPL), method of moments (MM), and minimum distance (MD). The MM as well as the MPL methods are considered as rank-based parameter estimation methods. On the other hand, to estimate $\boldsymbol{\theta}$, one can first estimate the marginal distributions F_X and F_Y either parametrically or nonparametrically.

Suppose that the copula as well as the marginal distributions in Sklar's result (5.1) are absolutely continuous. Hence, their corresponding densities exist and they are denoted by lower-case letters. Then, the corresponding joint log-likelihood is

$$\ell(\theta, \eta_X, \eta_Y) = \sum_{i=1}^{n} \ln\left[c_\theta\left\{F_{X,\eta_X}(X_i), F_{Y,\eta_Y}(Y_i)\right\}\right] + \sum_{i=1}^{n} \ln\left\{f_{X,\eta_X}(X_i) f_{Y,\eta_Y}(Y_i)\right\} \tag{5.23}$$

where η_X and η_Y are respectively the parameters of the marginal distributions F_X and F_Y. Maximizing this joint log-likelihood expression can be tedious and time-consuming, particularly in high dimensions. In the following, we briefly present the main estimation methods with some examples.

5.5.1 Inference functions for margins method

In order to avoid computation and time-consuming problems, the IFM maximizes expression (5.23) in two steps. In the first step, estimates of η_X and η_Y are obtained by maximizing the marginal log-likelihood functions

$$\ell(\eta_X) = \sum_{i=1}^{n} \ln\{f_{X,\eta_X}(X_i)\} \text{ and } \ell(\eta_Y) = \sum_{i=1}^{n} \ln\{f_{Y,\eta_Y}(Y_i)\} \quad (5.24)$$

In the second step, the obtained estimates $\hat{\eta}_X$ and $\hat{\eta}_Y$ of the marginal parameters from maximizing (5.24) are plugged into the first term of right-hand side of (5.23) to get a log-likelihood function for the copula parameter $\boldsymbol{\theta}$ as

$$\ell(\theta) = \sum_{i=1}^{n} \ln\left[c_\theta\{F_{X,\hat{\eta}_X}(X_i), F_{Y,\hat{\eta}_Y}(Y_i)\}\right] \quad (5.25)$$

The obtained estimator of $\boldsymbol{\theta}$ is consistent and asymptotically normal. However, this estimator is to some extent less efficient than those resulting from the full maximization of (5.23), except in the neighborhood of independence, as well as the estimator of θ can be severely biased if the marginal models are misspecified.

5.5.2 Maximum Pseudo-likelihood method

In order to obtain an estimator of the parameter $\boldsymbol{\theta}$ not affected by the margins (especially when they are misspecified), it is suggested to consider maximizing the log pseudo-likelihood given by

$$\ell(\theta) = \sum_{i=1}^{n} \ln\left[c_\theta(F_{n,X}(X_i), F_{n,Y}(Y_i))\right] = \sum_{i=1}^{n} \ln\left[c_\theta\left(\hat{U}_i, \hat{V}_i\right)\right] \quad (5.26)$$

This is the MPL method and this formulation is rank-based, where the pseudo-observations are given in (5.4). Under reasonable regularity conditions, the MPL estimator is consistent and asymptotic normal. Generally, the MPL estimator is less efficient than the full maximum likelihood estimator in the case of properly specified margins, except at independence. However, in oractice, the MPL method is shown to have good performance. It is widely used and applied to both one-parameter and multi-parameter copulas. See also Hofert et al. (2018) for more updated and detailed description of MPL properties.

5.5.3 Moment-based method

As shown in Section 5.4, several common bivariate copula families are formulated via a real-valued parameter θ which can be expressed as a relation of Kendall's tau ($\theta = h(\tau_K)$) or Spearman's rho ($\theta = g(\rho_S)$). Some examples are given in Table 5.3. A direct MM estimate θ_n of θ is given either by $\theta_n = h(\tau_n)$ or $\theta_n = g(\rho_{S,n})$ where τ_n and ρ_{nS} are the classical estimators of τ_K and ρ_S respectively (given in Section 5.4). As a simple example, we have for the case of Clayton copula $\tau_K = \theta/(\theta + 2)$ which leads to $\theta_n = 2\tau_n/(1 - \tau_n)$.

In general, the estimator from MM is less efficient than those from IFM or MPL methods. In addition, in some cases they need to be adjusted in order to respect the acceptable range of values for the parameter θ. In the simplest example of bivariate FGM copula, for instance, one has $\tau_K = 2\theta/9$ and $\rho_{S,n} = \theta/3$, so that $\theta_n = 9\tau_n/2$ and $\theta_n = 3\rho_{S,n}$. It is possible that the latter can fall outside the interval $[-1, 1]$ such as when $\tau_n = 0.25$ or $\rho_{S,n} = 0.4$. However, given their simplicity, and the good properties of $\rho_{S,n}$ and τ_n (rank-based, consistent, and asymptotic normality), MM estimators can be used as interesting starting values for numerical maximization of the (pseudo) likelihood.

5.5.4 Multi-parameter copula estimation

Even though the previous methods are conceptually valid for multi-parameter copulas, usually they are developed, compared, and evaluated for one-parameter copulas. The multi-parameter copulas are still not enough studied as their one-parameter counterparts. They are attracting some attention in the recent years in hydrology (e.g., Ben Nasr & Chebana, 2019; Salvadori & De Michele, 2010) as well as in statistics and other fields. The MM method is extended to multi-parameter copulas. Another method is based on the so-called multivariate L-moments recently proposed for multi-parameter copulas. It has some interesting features for the case of short sample sizes such as in HFA. Based on simulations, the multivariate L-moment method performs well in terms of bias and computation time with reasonable RMSE (root mean squared errors) compared to other estimation methods (Ben Nasr & Chebana, 2019). Multivariare L-moments are briefly presented in the Appendix.

Example

We consider here the dataset employed in Chapter 2 related to flood volume and peak series (station 02ED003, Ontario, Canada). The

TABLE 5.10 Parameter estimation with different methods and copulas for flood volume and peak series employed in Chapter 2 (station 02ED003, Ontario, Canada).

	MPL	IFM	MM (inverse tau)	MM (inverse rho)
Frank	2.764	2.736	3.201	2.899
Clayton	0.987	0.761	0.961	0.864
Gumbel	1.329	1.254	1.480	1.437
Joe	1.324	1.164	1.869	NA
Plackett	3.135	3.093	4.491	4.029
Normal	0.500	0.442	0.488	0.453

objective here is not to evaluate the performance of the different methods but rather to apply them to a given real-world dataset. The evaluation of these methods is the object of different simulation studies presented in the literature. Table 5.10 presents the obtained estimators with different methods for a number of copulas.

Overall, the considered estimation methods provided values in similar range for a given copula, with some exceptions. For instance, as exceptions, for Frank copula, we have MM inverse tau, whereas for Clayton copula we have IFM that are relatively different to the other estimators. However, for Gumbel and Plackett copulas, we have two categories: on the one hand likelihood-based methods (IFM and MPL) and on the other hand the MM method. Based on the literature performance assessment of these methods, it would be reasonable to select the MPL estimator.

In this example, it is important to mention that the IFM method requires known margins. Hence, the IFM method is performed on the GEV (Generalized Extreme Value) distribution for both margins. Note that for the FGM copula, both the MPL and IMF methods do not converge and MM does not exist. As found in Chapter 3, an outlier is detected for this dataset (corresponding to the year 1990). We considered here both results by including or excluding this outlier. We found here that all the methods are very slightly affected by the outlier for all considered copulas. Hence, for the simplicity of the presentation, the results excluding the outlier are not presented.

5.6 Copula selection

In multivariate HFA among other analysis, given a multivariate dataset, an important task is to select the most appropriate multivariate

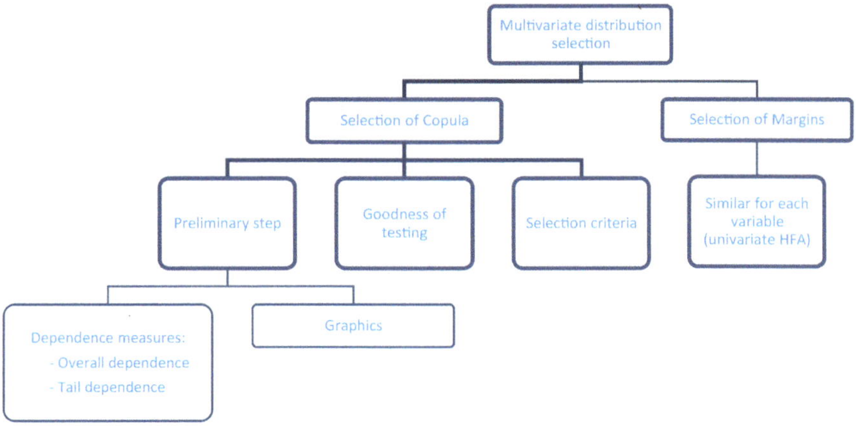

FIGURE 5.6 Overview of the main steps to select multivariate distribution.

distribution $F(.,.)$ for that data. According to Sklar result in (5.1), the selection of the joint distribution F is equivalent to the selection of a copula C and the margins F_X and F_Y. Here, the focus is on the selection of the copula, whereas the selection of the margins is briefly discussed at the end of this section.

Different copula models can be considered in order to find the best one in characterizing the dependence structure between variables. As any model, copula selection should follow several steps. The choice of the appropriate copula is important in HFA and it is not always an easy task. Copula selection should be progressive, as different issues need to be considered including visualization, dependence structure of the data, parameter estimation method, goodness-of-fit tests, selection criteria, and tail dependence. Fig. 5.6 summarizes the main parts of the selection of the joint multivariate distribution F where detailed descriptions are provided below.

5.6.1 Preliminary step

Before considering formal testing and selection criteria, it is useful to start this preliminary step. The latter can guide the selection in the sense that it helps, for instance, to exclude some copulas given their range of dependence or their shape.

Dependence measures: The dependence structure of the data needs to be analyzed to initially select potential copula candidates. First, we consider the most widespread quantitative measures which are the rank-based nonparametric measures: Kendall's τ_K and Spearman's ρ_s

(see Section 5.4). These measures are used in order to cut down the number of potential copulas, since not all copulas support all τ_K or ρ_s values as shown in Section 5.4. As an example, the Ali–Mikhail–Haq copula should be rejected for a sample with empirical $\tau_K = 0.6$ since the corresponding τ_K to that copula should be in the range $[-0.1817, 0.3333]$.

Moreover, the tail dependence should be also taken into account in the copula selection process since in HFA, and other applications, the interest is on extremes. A number of studies have stressed that ignoring UTD could lead to less accurate estimation of flood risk (e.g., Requena et al., 2013). The tail dependence measures are presented in Section 5.4. For a given multivariate sample, one can evaluate the sample tail dependence measures and then compare them with those of potential copulas. This allows to discard some copulas if the sample tail dependence value is very different to the one of a given copula (see Table 5.8).

Example

From the previous example, based on the dataset used in Chapter 2 for flood volume V and flood peak Q (station 02ED003, Ontario, Canada), the FGM copula should be discarded for further analysis on copula selection. Indeed, the values of τ_K or ρ_s coefficients are both outside of their respective ranges for this copula.

In terms of UTD, for this dataset, as shown in previous example, we have an empirical nonparametric measure $\hat{\lambda}_U^{CFG} = 0.3142$ in which case Archimedean (except Joe), Normal, and Plackett copulas can be excluded in the copula selection procedure in the case where the study aim is to focus on risk assessment. In particular, EV copulas (Gumbel, Galambos, Hüsler–Reiss, and Tawn) as well as Joe seem to be appropriate in this case. Even though their respective UTD are in the same range and close to $\hat{\lambda}_U^{CFG}$, only the one corresponding to Joe copula is relatively higher than the rest of the values which are all close to 0.4.

Graphics: Different graphical tools are available in order to provide interesting information about the dependence structure and hence help and guide selecting the appropriate copula through some characteristics such as the strength of dependence, symmetry, and behavior in the tail.

As a first and natural graphic is the scatter plot of the original data. However, in order to focus on copula, it is more appropriate to consider scatter plot based on the pseudo-observations given in (5.4) (see Section 5.2). In this case, it is called *rank* plot. Such a plot could provide many hints as the type of the dependence embodied in a copula C, guidance in selecting a parametric form for C, or confirm or not that the underlying copula C is different from the independence copula.

The pseudo-observations can also be transformed to the normal scale by setting $\hat{O}_i = (\Phi^{-1}(\hat{U}_i), \Phi^{-1}(\hat{V}_i))$ for $i = 1, \ldots, n$, where Φ is the cumulative distribution function of standard normal distibution. The obtained plot composed by $\hat{O}_1, \ldots, \hat{O}_n$ is called a *rankit* plot. Note that the obtained scatter plot could highlight departures from bivariate normality where a scatter plot of the later should look "elliptically contoured" (as in Fig. 5.2). In addition, in the scatter plot of normal scores, sharper corners indicate the presence of tail dependence (for more details and examples, see e.g., Joe, 2014; Sections 1.4 and 2.13). Finally, this plot is useful to visually detect nonexchangeability or radial symmetry of C, where a copula is radially symmetric if its density is symmetric with respect to the point (0.5, 0.5) and it is exchangeable if $C(u_1, u_2) = C(u_2, u_1)$ for all $u_1, u_2 \in [0, 1]$. In this regard, Hofert et al. (2018) provided some interesting examples.

In high dimensions, one can consider pairs of components of the above plots. However, in that case, rank-based plots provide only partial information on the overall dependence structure (e.g., Genest & Chebana, 2017). It is important to mention that the above plots (rank and rankit plots) can be less useful when the sample size is large. In HFA, this can happen for instance in regional HFA where the data from a number of stations from a given region are pooled (see Chapter 8).

The so-called K-plot can provide additional insight to the previous plots. It consists in plotting the real-valued points $\hat{W}_i = \hat{C}_n(\hat{U}_i, \hat{V}_i)$, $i = 1, \ldots, n$. As previously indicated (Section 3.1), the empirical distribution function $\hat{K}_n$ associated with the set $\hat{W}_1, \ldots, \hat{W}_n$ is an estimate of the distribution function $K(t) = P\,[C(U, V) \leq t]$, $t \in [0, 1]$ of the variable $W = C(U, V)$, which can be obtained given choices of C. In the particular case of independence, we have $K(w) = w - w\ln(w)$ for all $w \in (0, 1)$. This is useful to check if it is appropriate to assume independence based on visual comparison between $\hat{K}_n$ and K or by drawing the corresponding QQ plot (Quantile-Quantile plot). More details and descriptions for high-dimension can be found, for instance, in Genest and Boies (2003).

Example

We consider the dataset employed in Chapter 2 regarding flood volume and peak (station 02ED003, Ontario, Canada). In addition, in Chapter 4, we found that this dataset satisfies the basic assumptions (stationarity, homogeneity, and serial independence). Hence, it is appropriate to continue the analysis with this dataset.

Fig. 5.7 presents different plots for this dataset as well as the contour plots for a number of copulas. From the rank-plot, we observe positive dependence and hence departure from independence. In addition, the rankit plot suggests a departure from normality since it does not seem to be elliptical but slightly symmetric. However, we cannot

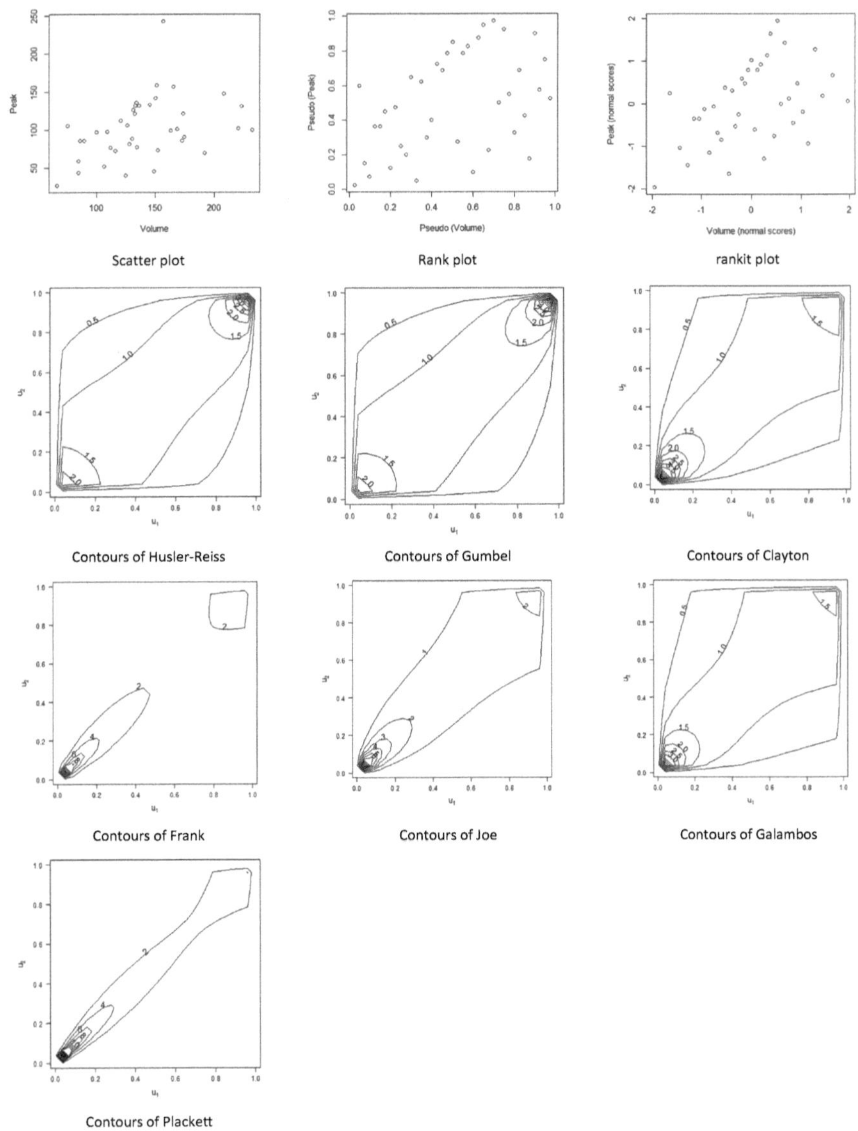

FIGURE 5.7 Different plots for the flood volume and peak data set employed in Chapter 2 (station 02ED003, Ontario, Canada).

conclude the existence of UTD since the scatter plot is not sharp in the upper corner. The lack of symmetry with respect to the point (0.5, 0.5) also suggests that the appropriate copula is not radially symmetric. Based on the different density contours, we can observe that some

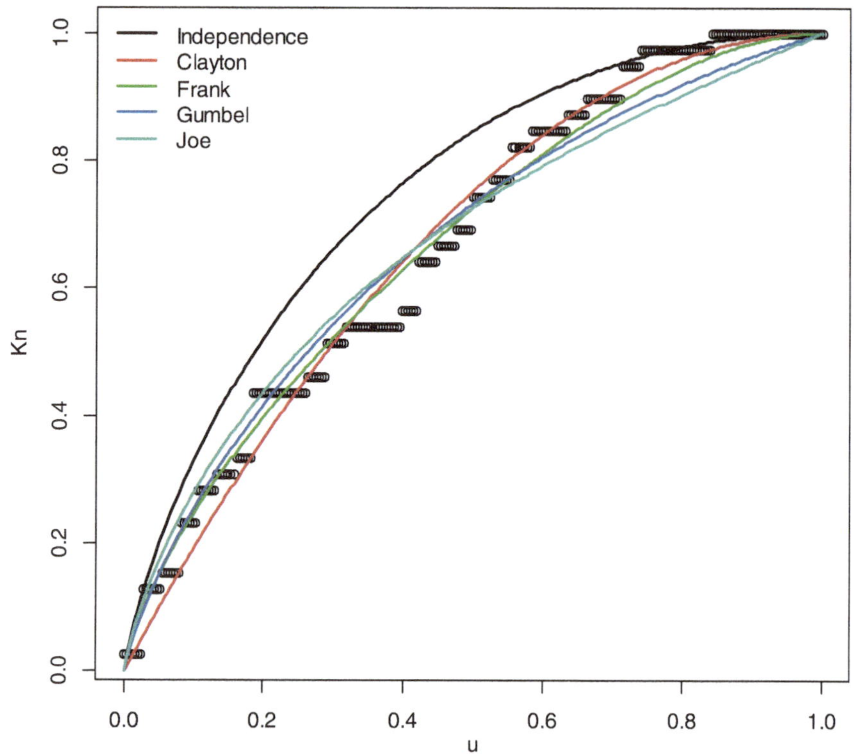

FIGURE 5.8 K function for different copulas (with tau from the data) and empirical plot function for data in Chapter 2 (station 02ED003, Ontario, Canada).

copulas could be more appropriate than others. Indeed, Hüsler–Reiss, Gumbel, Clayton, and Galambos copulas could be good choices. From the K-plot (Fig. 5.8), it seems that the Archimedean copulas fail to fit the upper tail as it can be seen on the second half of the range. However, based on the first half of the range, the Frank copula could be better than the others.

5.6.2 Copula goodness-of-fit testing

After the preliminary step, it is important and necessary to proceed with a formal goodness-of-fit testing. First, before going further in the analysis, it is important to test if the copula C is the independence one. If it is the case, then it is enough to focus only on marginal modeling as in the classical univariate HFA (see also the end of this section). Otherwise, the null hypothesis to be tested becomes H_0: $C \in C_\theta$ for a given parametric copula family C_θ. This is the aim of goodness-of-fit tests for copula models.

A number of goodness-of-fit tests for copulas were proposed as reviewed by a number of studies. According to the comparisons based on extensive simulations, the test based on the empirical copula $\hat{C}_n$, a consistent estimator of the unknown copula C, seems to perform particularly well. Hence, a natural goodness-of-fit test is based on the deviation between $\hat{C}_n$ and $C_{\theta n}$, the latter being an estimator of C under the null hypothesis where $\boldsymbol{\theta}_n$ is an estimator of $\boldsymbol{\theta}$ obtained from the pseudo-observations (such as the MPL estimator) or any other consistent estimator (see Section 5.5). This test is a part of the so-called *blanket tests* for which no choice of tuning parameters is required such as smoothing parameter, weight function, kernel, or window. Other goodness-of-fit tests have been also proposed such as those based on the Rosenblatt transformation or those based on nonparametric estimators of the copula density. The above tests are general and conceptually valid for any copula. However, specific tests have been also developed for particular dependence structures such as for Gaussian or Clayton copulas, or even EV and Archimedean copula classes. The reader is referred for instance to Hofert et al. (2018) for more details and more specific references. Goodness-of-fit tests for a large class C of copulas are also attracting attention recently, such as the class of bivariate exchangeable copulas, spatial copulas (Quessy & Durocher, 2019), and multi-parameter copulas (Ben Nasr & Chebana, 2019).

As introduced earlier, the most powerful version of this procedure is based on the Cramér–von Mises statistic

$$S_n^{\mathrm{GoF}} = \sum_{i=1}^{n} \left\{ \hat{C}_n\left(\hat{U}_i, \hat{V}_i\right) - C_{\theta_n}\left(\hat{U}_i, \hat{V}_i\right) \right\}^2 \tag{5.27}$$

Given the complexity of this test statistic, the corresponding p-value can be approximated with a parametric bootstrap. Hence, a copula family will be accepted if the associated p-value is larger than the common $\alpha_{\mathrm{level}} = 0.05$. Note that the p-value can be only employed to accept or reject a copula family, but not as a criterion to provide a ranking of the accepted copulas.

The Cramér–von Mises test has some interesting properties. Indeed, it is consistent for any copula family and is easier to calculate. In the bivariate case with one-parameter copulas, faster goodness-of-fit tests are often obtained when using MM estimator instead of MPL estimator. Test statistics based on the deviation between the empirical copula and the copula with the estimated parameter can be defined with other distances such as the L^1 and L^∞ distances (L^2 corresponds to Cramer–von Mises and L^∞ to Kolmogorov–Smirnov statistics). However, these tests are generally less powerful.

Given the importance of the Archimedean and EV copula classes in HFA and other applications, it is worth to briefly present the corresponding goodness-of-fit tests. Similar to (5.27), another test can be defined based on an L^2 deviation between the empirical Kendall distribution K and its parametric estimator $K_{\theta n}$ under the null hypothesis:

$$T_n = \sum_{i=1}^{n} \left\{ \hat{K}_n\left(\hat{W}_i\right) - K_{\theta_n}\left(\hat{W}_i\right) \right\}^2 \tag{5.28}$$

where K_{θ_n},$\hat{K}_n$ and $\hat{W}_i = \hat{C}_n(\hat{U}_i, \hat{V}_i)$ are given in Section 5.3.1. Since K is better to characterize Archimedean copulas, this test should perform best with Archimedean copula families. Its p-value is approximated similarly as for the previous Cramer–von Mises test (5.27).

In order to check if an EV copula is appropriate to a given dataset (testing of extremeness), one should first test the hypothesis H_E: $C \in E_2$ where E_2 denotes the class of bivariate EV copulas. To this end, the following test statistic is considered

$$S_n^{\mathrm{GKR}} = -1 + \frac{8}{n(n-1)} \sum_{i \neq j}^{n} I_{ij} - \frac{9}{n(n-1)(n-2)} \sum_{i \neq j \neq k}^{n} I_{ij} I_{kj} \tag{5.29}$$

where $I_{ij} = 1(X_j \le X_i, Y_j \le Y_i)$ for all $i, j = 1, \ldots, n$. Under H_E, this statistic is asymptotically normal with zero mean with an estimated variance, for instance by jackknife. The performance of this test is more than acceptable considering a large variety of alternatives.

After applying the previous extremeness test and if it reveals no evidence against H_E, then as a next step it is of interest to check the appropriateness of a specific given EV copula family for the data at hand. As previously indicated (Section 5.3.2), an EV copula is characterized by the Pickands dependence function A. Hence, the null hypothesis becomes H_A: $A \in \mathrm{A}$. Then, it is natural to define the statistics based on the deviation between the best representative $A_{\theta n}$ from the class A to a model-free, nonparametric estimator A_n (the estimators given in Eqs. 5.11 and 5.12). Formally, in the bivariate case, the proposed tests are based on the statistics

$$T_n^{\mathrm{GKNY1}} = \int_0^1 \left\{ \hat{A}_n^{\mathrm{CFG}}(t) - A_{\theta_n}(t) \right\}^2 dt \text{ and } T_n^{\mathrm{GKNY2}} = \int_0^1 \left\{ \hat{A}_n^{P}(t) - A_{\theta_n}(t) \right\}^2 dt \tag{5.30}$$

The corresponding p-values are approximated similarly as for the previous Cramer–von Mises test (5.27). Based on a simulation study, the statistic T_n^{GKNY1} typically leads to a more powerful test than the statistic T_n^{GKNY2} as well as than other general tests.

The p-value approximation of all the above tests by parametric bootstrap has its high computational cost, especially when the sample size increases. However, to overcome this limitation, a faster and large-sample testing procedure has been developed. It is based on *multiplier* central limit theorems. Monte Carlo simulations, with some values of the dimension d and the sample size and under H_0, showed that both resampling procedures are asymptotically equivalent. However, this result is not necessary valid under the alternative (Hofert et al., 2018).

In practice, for a given dataset, goodness-of-fit tests should be applied to a (large) number of potential candidates of models (here copulas). Usually, they result to a subgroup of accepted models. Note that when applying goodness-of-fit tests, some particular cases could arise where for small sample size, none of the candidate copulas are rejected (all potential copulas are plausible), whereas for high dimension or large sample size, it is possible that all the copula candidate families are rejected.

Example

As in the previous sections, we consider the dataset of flood volume and peak employed in Chapter 2 (station 02ED003, Ontario, Canada). First, we considered the test of extremness (5.29) where the obtained p-value is 0.02679 (with a statistics value 2.2146). Hence, if we consider $\alpha = 1\%$, we can accept the null hypothesis that the data have extreme dependence. Subsequently, it is reasonable to test for specific EV copulas with the tests T_n^{GKNY1} and T_n^{GKNY2} in (5.30). The obtained results are given in Table 5.11. We observe that both tests suggest to accept each of the EV copulas (Gumbel, Tawn, Galambos, and Hüsler–Reiss) at $\alpha = 5\%$. This result is consistent with the one obtained with the previous test of extremness. Even though for a given copula the p-value corresponding to T_n^{GKNY1} is higher than the one of T_n^{GKNY2}, the result in terms of decision remains the same.

TABLE 5.11 Goodness-of-fit testing results for extremness for the dataset in Chapter 2 for flood volume and peak (station 02ED003, Ontario, Canada).

		T_n^{GKNY1}		T_n^{GKNY2}	
	θ_n	**Statistic**	***p*-value**	**Statistic**	***p*-value**
Gumbel	1.3288	0.0201	**0.3631**	0.10304	**0.0684**
Tawn	0.6433	0.0231	**0.4760**	0.10303	**0.0924**
Galambos	0.6185	0.0194	**0.2672**	0.08452	**0.1084**
Hüsler–Reiss	1.0499	0.0221	**0.1703**	0.06562	**0.1533**

Bold character indicates the corresponding copula can be accepted at 5%.

TABLE 5.12 Goodness-of-fit testing results for the dataset in Chapter 2 for flood volume and peak (station 02ED003, Ontario, Canada).

			***p*-value**	
	θ_n	**Statistic**	**Parametric bootstrap**	**multiplier**
Frank	2.7643	0.0430	**0.0994**	**0.0265**
Clayton	0.9874	0.0381	NC	**0.0395**
Joe	1.3242	0.1010	NC	0.0005
Gumbel	1.3288	0.0590	NC	**0.0185**
Tawn	0.6433	0.0595	NC	0.0075
Galambos	0.6185	0.0543	NC	**0.0105**
Hüsler–Reiss	1.0499	0.0493	NC	**0.0175**
Normal	0.5004	0.0315	**0.2323**	**0.0994**
Plackett	3.1353	0.0553	0.0145	**0.0125**

Bold character indicates the corresponding copula can be accepted at 1%. NC means does not converge.

It is also important to test for other copulas using the Cramér–on Mises test in (5.27). The obtained results are given in Table 5.12. To evaluate the *p*-values, we considered the parametric bootstrap as well as the multiplier algorithm. We observe that the parametric bootstrap does not converge for the majority of the copulas. However, the multiplier algorithm was very fast and converges for all situations. By choosing $\alpha = 1\%$, Tawn and Joe copulas are rejected. Note that the test cannot be applied for both the FGM and AMH copulas since the Kendall's tau is out of their respective range.

We also considered the situations of including and excluding the detected outlier corresponding to year 1990 (Chapter 3). However, the obtained results are very similar and especially lead to the same conclusions. Hence, we presented only the results for all data including that outlier.

5.6.3 Selection criteria for copula

Once at least a goodness-of-fit test has been applied and led to identify (acceptance) a number of potential candidate copulas, several criteria can be used to help selecting the most appropriate copula by providing a ranking of those accepted ones. An ideal situation is when the selected family is among those accepted by a goodness-of-fit. This is not always the case in some applications. In such a situation, it is suggested to enlarge the set of the copula families to be tested initially.

Copula selection criteria have been already used in several studies and in particular in hydrology (e.g., Genest & Chebana, 2017; Klein et al., 2010; Requena et al., 2013). Criteria such as the AIC_{cop} and the BIC_{cop}, adapted version to the copula modeling of the well-known Akaike Information Criterion (AIC) and Bayesian Information Criterion (BIC) respectuvely, can be used to select among candidate copula models with different number of parameters:

$$AIC_{cop} = -2\sum_{i=1}^{n} \ln\left[c\left(\frac{R_i}{n+1}, \frac{S_i}{n+1}; \theta\right)\right] + 2k_{cop};\ BIC_{cop} = AIC_{cop} - (2 - \ln(n))k_{cop} \quad (5.31)$$

where $c(u, v; \hat{\theta})$ is the copula density function and k_{cop} is the corresponding number of parameters of C. The best copula is the one corresponding to the minimum value of the considered criterion. Commonly, the AIC_{cop} definition is based on the maximum likelihood (ML) of a given model. However, for copula parameters, the MPL is preferred as an estimation method (see Section 5.5). Hence, the AIC_{cop} based on MPL becomes more accurate than the one directly based on ML (similarly for BIC_{cop}). The corresponding AIC expression becomes (similarly for BIC_{cop}):

$$AIC_{cop} = -2\log L_n(\hat{\theta}_n) + 2\,k_{cop} \quad (5.32)$$

where $L_n(\hat{\theta}_n)$ is the value of the pseudo-likelihood function $L_n(\theta)$ at the MPL estimator $\hat{\theta}_n$. Several studies considered this formulation. For theoretical developments related to model selection for copulas, see Grønneberg and Hjort (2014).

Beside AIC_{cop} and BIC_{cop}, other criteria have been also proposed in the literature. For instance, a copula information criterion (CIC) has been developed for model selection with maximum likelihood estimation in two stages. In addition, we have the cross-validation copula information criterion. A comparison with AIC_{cop} (5.32) in a bivariate simulation study concluded that overall the difference between the two selection criteria is not significant and hence it is suggested to keep using the usual AIC_{cop} (5.32). In addition, the cross-validation is not appropriate for the HFA framework since the aim in the latter is not the forecasting (prediction) but risk assessment. The topic of selection criteria for copula seems to be recent and evolving (see e.g., Ko & Hjort, 2019).

Example

As a continuation of the examples of the previous sections, we consider the same dataset, that is, flood volume and peak series employed in Chapter 2 (station 02ED003, Ontario, Canada). The criteria AIC_{cop}

TABLE 5.13 AIC_{cop} and BIC_{cop} for different copulas for the dataset of Chapter 2 (station 02ED003, Ontario, Canada).

	AIC_{cop}	BIC_{cop}
Clayton[a]	**− 9.6477**	**− 7.9841**
Normal	− 6.8136	− 5.1501
Frank	− 5.4420	− 3.7785
Plackett	− 4.5556	− 2.8920
Hüsler−Reiss[b]	*− 3.9031*	*− 2.2395*
Galambos	− 3.4196	− 1.7561
Gumbel	− 2.6306	− 0.9671
Tawn	− 1.2264	0.4372
Joe	0.1588	1.8223

[a]*The bold character indicates the smallest value of the criterion.*
[b]*The italic character indicates the smallest value of the criterion among the class of extreme value copulas.*

and BIC_{cop} are evaluated for all copulas considered in Table 5.12. The obtained results are presented in Table 5.13. First, note that the rejected copulas Joe and Tawn (Table 5.12) are indeed the last ones in the list according to AIC_{cop} and BIC_{cop}. This result indicates an agreement between the goodness-of-fit testing and the model selection procedures.

The smallest values of AIC_{cop} and BIC_{cop} are associated to the Clayton copula. It was accepted by the Cramer−von Mises goodness-of-fit test (multiplier) with p-value = 0.0395. Hence, we can select the Clayton copula as an appropriate model for overall dependence structure. However, if we consider the tail dependence, which is usually more appropriate in the HFA framework, it is the Husler−Reiss copula that should be selected based on the AIC_{cop} and BIC_{cop} values, but also the previous analysis including the extremness tests, the Cramer−von Mises goodness-of-fit test, as well as the preliminary analysis (UTD and graphics).

5.6.4 Margin modeling

Once the most appropriate copula is chosen, then for a complete multivariate HFA, from Sklar's theorem in (5.1), it remains to select the appropriate margins F_X and F_Y (Fig. 5.6). The main steps of copula selection procedure remain valid although with different statistical tools. We briefly present the modeling step for the univariate setting (parameter estimation, goodness-of-fit testing, and selection criteria) below.

Different univariate distributions with widespread use in HFA can be considered as potential candidates for fitting the studied variables in the univariate setting. For instance, Gumbel (G), GEV, two-parameter log-Normal (LN2), Pearson type III (P3), and log-Pearson type III (LP3) are among the most employed distributions. A number of methods have been developed to estimate the parameters of univariate distributions, such as MM, method of maximum likelihood, method of L-moments, as well as their extensions and different versions (general or specific for a given distribution). Formulations and discussions regarding these univariate distributions can be found in a number of textbooks or chapters such as Rao and Hamed (2000), or recently Singh (2017, Chapter 21) with a very wide list of univariate distributions with discussion.

The choice of the adequate distribution is determined on the basis of numerous classical and recent statistical tools, including graphical representations (probability plots and QQ-plots) and goodness-of-fit tests such as the tests of Chi-squared, Kolmogorov–Smirnov, and Cramer–von Mises. These tests are widely used in the univariate HFA literature. Goodness-of-fit tests based on the empirical distribution function $F_n(x)$ mainly include Kolmogorov–Smirnov (K-S), Cramer–von Mises, and Anderson–Darling tests given respectively by

$$D_n = \sup_{x \in \mathbb{R}} |F_n(x) - F(x)| \tag{5.33}$$

$$W_n{}^2 = \int_{-\infty}^{\infty} [F_n(x) - F(x)]^2 dF(x) \tag{5.34}$$

$$A_n = n \int_{-\infty}^{\infty} \frac{(F_n(x) - F(x))^2}{F(x)(1 - F(x))} dF(x) \tag{5.35}$$

The p-value of each of the above statistics is usually computed based on the corresponding limiting probability distributions. However, as for copula goodness-of-fit tests, the parametric bootstrap method can be considered to obtain the corresponding p-values as well. Another well-known goodness-of-fit test different to the previous ones is the Chi-square. It is not based on the empirical distribution with a statistic given by

$$\chi_n^2 = \sum_{j=0}^{k} \frac{(o_j - e_j)^2}{e_j} \tag{5.36}$$

where o_j is the observed frequency count for the level-j of a variable, e_j is the corresponding expected frequency count from the fitted probability distribution, and k is the number of levels of the random variable. This test rejects the null hypothesis if χ_n^2 is larger than the quantile of order $(1-\alpha)$ of Chi-square distribution with $k-1$ degrees of freedom. Laio et al. (2009) performed a comparison of several goodness-of-fit tests for marginal distribution selection. They recommended a transformation of the Anderson–Darling test for hydrological analysis.

After goodness-of-fit test application, the selection of the best distribution among the set that has passed the test can be done via model selection criteria. In the present study, the two well-known criteria AIC and BIC are considered. The best distribution for each criterion is the one that minimizes AIC_{mar} and the BIC_{mar}, respectively (the same holds for Y):

$$AIC_{mar} = -2\ln\left(\prod_{i=1}^{n} \hat{f}_X(x_i, \hat{\eta}_X)\right) + 2k_{mar,X} \tag{5.37}$$

$$BIC_{mar} = -2\ln\left(\prod_{i=1}^{n} \hat{f}_X(x_i, \hat{\eta}_X)\right) + \ln(n)k_{mar,X} \tag{5.38}$$

where $\hat{f}_X(x_i, \hat{\eta}_X)$ is the estimated marginal density distribution of X, $\hat{\eta}_X$ is the estimated parameter, and $k_{mar,X}$ is the number of parameters of the distribution. Laio et al. (2009) checked the behavior of several model selection criteria for hydrological data. Their study showed that no criterion could be considered better than the others.

Example

We continue to use the same dataset as in previous sections, regarding flood volume and peak (employed also in Chapter 2, station 02ED003, Ontario, Canada). First, we note here in the univariate framework a significant difference when excluding the detected outlier on the modeling of the peak (but not the volume) as also noticed in Chapter 3 (which was not the case for copula selection). Since our focus is on the dependence structure and copula modeling, the

TABLE 5.14 Summary of the modeling results of flood peak and volume of the dataset of Chapter 2 (station 02ED003, Ontario, Canada).

	Variable	Selected distribution	Estimated parameters	
All data				
	Peak	Gamma	$\alpha = 0.062$	$\lambda = 6.27$
	Volume	Gamma	$\alpha = 0.085$	$\lambda = 11.92$
Excluding outlier 1990				
	Peak	Weibull	$\alpha = 109.02$	$k = 3.33$
	Volume	Gamma	$\alpha = 0.084$	$\lambda = 11.67$

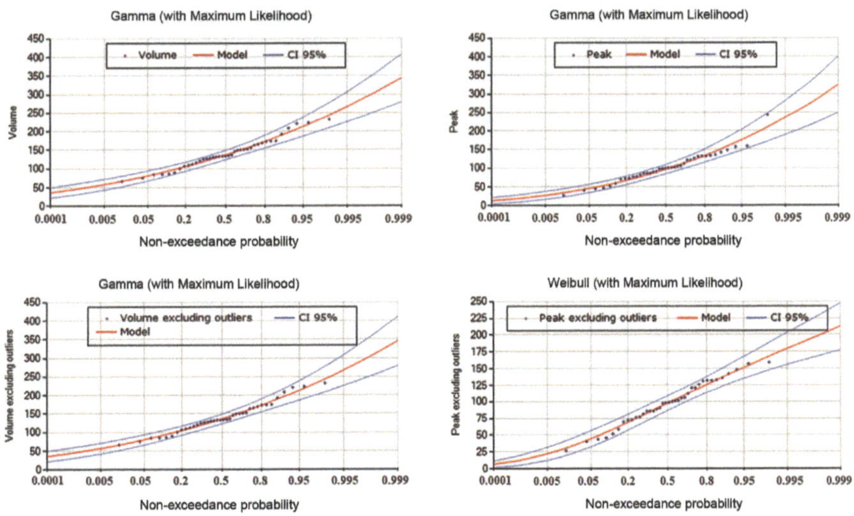

FIGURE 5.9 Nonexceedance probability curves of the selected distributions for each case of flood volume and peak of the dataset employed in Chapter 2 (station 02ED003, Ontario, Canada).

TABLE 5.15 AIC and BIC criteria for the top three distributions for each case of flood volume and peak of the dataset employed in Chapter 2 (station 02ED003, Ontario, Canada).

	Distribution	AIC	BIC
Volume (all data)	**Gamma**	**401.30**	**404.62**
	Log Normal 2	402.02	405.34
	Gumbel	402.72	406.05
Peak (all data)	**Gamma**	**399.11**	**402.44**
	Gumbel	399.71	403.03
	GEV	400.92	405.91
Volume (excluding 1990)	**Gamma**	**391.65**	**394.92**
	Log Normal 2	392.22	395.49
	Gumbel	392.83	396.10
Peak (excluding 1990)	**Weibull**	**376.80**	**380.08**
	Log Pearson	377.39	382.30
	Normal	377.52	380.79

The bold character indicates the smallest value of the criterion (and hence the selected marginal distribution).

univariate results are briefly presented. The obtained results of the selected distributions are summarized in Table 5.14.

In order to avoid confusion regarding the parameterization of the distributions in Table 5.14, recall that the densities of Gamma and Weibull are given respectively by:

$$f_{\text{Gamma}}(x) = \frac{\alpha}{\Gamma(\lambda)} e^{-\alpha x} (\alpha x)^{\lambda - 1}, \quad \alpha > 0, \lambda > 0, x > 0 \tag{5.39}$$

$$f_{\text{Weibull}}(x) = \frac{k}{\alpha} \left(\frac{x}{\alpha}\right)^{k-1} e^{-x/\alpha}, \quad \alpha > 0, k > 0, x > 0 \tag{5.40}$$

In Fig. 5.9, we present the nonexceedance probability curves for each of the selected distributions (from Table 5.14). We can observe that each selected distribution fits well the corresponding data. Formally, in Table 5.15, we present the AIC and BIC criteria of the three top distributions (even though a larger number of distributions is treated). These values confirm the selected distributions.

References

Ayantobo, O. O., Li, Y., & Song, S. (2019). Multivariate drought frequency analysis using four-variate symmetric and asymmetric Archimedean copula functions. *Water Resources Management, 33*(1), 103–127.

Ben Alaya, M., Zwiers, F., & Zhang, X. (2018). Probable maximum precipitation: Its estimation and uncertainty quantification using bivariate extreme value analysis. *Journal of Hydrometeorology, 19*(4), 679–694.

Ben Nasr, I., & Chebana, F. (2019). Multivariate L-moment based tests for copula selection, with hydrometeorological applications. *Journal of Hydrology, 579*, 124151.

Chebana, F. (2013). Multivariate analysis of hydrological variables. In A.-H. El-Shaarawi, & W. Piegorsch (Eds.), *Encyclopedia of environmetrics* (2nd ed., pp. 1676–1681). Chichester, UK: John Wiley & Sons Ltd. Available from 10.1002/9780470057339.vnn044.

Chebana, F., & Ouarda, T. B. (2011). Multivariate quantiles in hydrological frequency analysis. *Environmetrics (London, Ont.), 22*(1), 63–78.

Chen, L., & Guo, S. (2019). *Copulas and its application in hydrology and water resources*. Springer.

Czado, C. (2019). *Analyzing dependent data with vine copulas. Lecture Notes in Statistics*. Springer.

Genest, C., & Boies, J.-C. (2003). Detecting dependence with Kendall plots. *The American Statistician, 57*, 275–284.

Genest, C., & Chebana, F. (2017). In V. P. Singh (Ed.), *Handbook of applied hydrology* (2nd ed., pp. 301–309). New York: McGraw-Hill. (Chapter 30).

Genest, C., & Nešlehová, J. (2012b). In (2nd ed.A. H. El-Shaarawi, & W. W. Piegorsch (Eds.), *Copulas and copula models. Encyclopedia of environmetrics* (Vol. 2Chichester: Wiley.

Grimaldi, S., & Serinaldi, F. (2006). Design hyetograph analysis with 3-copula function. *Hydrological Sciences Journal, 51*(2), 223–238.

Grønneberg, S., & Hjort, N. L. (2014). The copula information criteria. *Scandinavian Journal of Statistics, 41*(2), 436–459.

Hao, Z., & Singh, V. P. (2016). Review of dependence modeling in hydrology and water resources. *Progress in Physical Geography, 40*(4), 549–578.

Hofert, M., Kojadinovic, I., Mächler, M., & Yan, J. (2018). *Elements of copula modeling with R*. Springer.

Joe, H. (2014). *Dependence modeling with copulas*. CRC Press.

Kao, S. C., & Govindaraju, R. S. (2007). A bivariate frequency analysis of extreme rainfall with implications for design. *Journal of Geophysical Research: Atmospheres, 112*(D13).

Karahacane, H., Meddi, M., Chebana, F., & Saaed, H. A. (2020). Complete multivariate flood frequency analysis, applied to northern Algeria. *Journal of Flood Risk Management, 13*(4), e12619.

Klein, B., Pahlow, M., Hundecha, Y., & Schumann, A. (2010). Probability analysis of hydrological loads for the design of flood control systems using copulas. *Journal of Hydrologic Engineering-ASCE, 15*, 360–369.

Ko, V., & Hjort, N. L. (2019). Copula information criterion for model selection with two-stage maximum likelihood estimation. *Econometrics and Statistics, 12*, 167–180.

Laio, F., Di Baldassarre, G., & Montanari, A. (2009). Model selection techniques for the frequency analysis of hydrological extremes. *Water Resources Research, 45*(7).

Lekina, A., Chebana, F., & Ouarda, T. B. (2015). On the tail dependence in bivariate hydrological frequency analysis. *Dependence Modeling, 3*(1).

Li, F., & Zheng, Q. (2016). Probabilistic modelling of flood events using the entropy copula. *Advances in Water Resources, 97*, 233–240.

Liu, Y., Li, Y., Ma, Y., Jia, Q., & Su, Y. (2020). Development of a Bayesian-copula-based frequency analysis method for hydrological risk assessment–The Naryn River in Central Asia. *Journal of Hydrology, 580*, 124349.

Ma, M., Song, S., Ren, L., Jiang, S., & Song, J. (2013). Multivariate drought characteristics using trivariate Gaussian and Student t copulas. *Hydrological Processes, 27*(8), 1175–1190.

Mikosch, T. (2006). Copulas: Tales and facts. *Extremes, 9*(1), 3–20.

Nelsen, R. B. (2006). *An introduction to copula* (2nd ed.). New York: Springer.

Papaioannou, G., Kohnová, S., Bacigál, T., Szolgay, J., Hlavèová, K., & Loukas, A. (2016). Joint modelling of flood peaks and volumes: A copula application for the Danube River. *Journal of Hydrology and Hydromechanics, 64*(4), 382.

Piantadosi, J., Howlett, P., Borwein, J., & Henstridge, J. (2012). Maximum entropy methods for generating simulated rainfall. *Numerical Algebra, Control & Optimization, 2*(2), 233.

Quessy, J. F., & Durocher, M. (2019). The class of copulas arising from squared distributions: Properties and inference. *Econometrics and Statistics, 12*, 148–166.

Rao, A. R., & Hamed, K. H. (2000). *Flood frequency analysis*. Boca Raton: CRC Press.

Requena, A., Mediero, L., & Garrote, L. (2013). A bivariate return period based on copulas for hydrologic dam design: Accounting for reservoir routing in risk estimation. *Hydrology and Earth System Sciences, 17*(8), 3023–3038.

Salvadori, G., & De Michele, C. (2004). Frequency analysis via copulas: Theoretical aspects and applications to hydrological events. *Water Resources Research, 40*(12).

Salvadori, G., & De Michele, C. (2010). Multivariate multiparameter extreme value models and return periods: A copula approach. *Water Resources Research, 46*(10).

Salvadori, G., & De Michele, C. D. (2011). Estimating strategies for multiparameter multivariate extreme value copulas. *Hydrology and Earth System Sciences, 15*(1), 141–150.

Salvadori, G., De Michele, C., Kottegoda, N. T., & Rosso, R. (2007). *Extremes in nature: An approach using copulas* (Vol. 56). Springer Science & Business Media.

Serinaldi, F. (2008). Analysis of inter-gauge dependence by Kendall's τ K, upper tail dependence coefficient, and 2-copulas with application to rainfall fields. *Stochastic Environmental Research and Risk Assessment, 22*(6), 671–688.

Serinaldi, F., & Grimaldi, S. (2007). Fully nested 3-copula: Procedure and application on hydrological data. *Journal of Hydrologic Engineering, 12*(4), 420–430.

Shafaei, M., Fakheri-Fard, A., Dinpashoh, Y., Mirabbasi, R., & De Michele, C. (2017). Modeling flood event characteristics using D-vine structures. *Theoretical and Applied Climatology, 130*(3), 713–724.

Sharma, S., & Mujumdar, P. (2019). On the relationship of daily rainfall extremes and local mean temperature. *Journal of Hydrology, 572*, 179–191.

Singh, V. P. (2017). *Frequency distributions, chapter 21. Handbook of applied hydrologyedition* (2nd ed.). New York: McGraw-Hill.

Sklar, M. (1959). *Fonctions de repartition an dimensions et leurs marges* (8, pp. 229–231). Paris: Institut de Statistiques de l'Université de.

Song, S., & Singh, V. P. (2010). Meta-elliptical copulas for drought frequency analysis of periodic hydrologic data. *Stochastic Environmental Research and Risk Assessment, 24*(3), 425–444.

Tosunoglu, F., & Singh, V. P. (2018). Multivariate modeling of annual instantaneous maximum flows using copulas. *Journal of Hydrologic Engineering, 23*(3)04018003.

Trivedi, P. K., & Zimmer, D. M. (2007). *Copula modeling: An introduction for practitioners*. Now Publishers Inc.

Zhang, L., & Singh, V. P. (2019). *Copulas and their applications in water resources engineering*. Cambridge University Press.

Zhu, Y., Liu, Y., Wang, W., Singh, V. P., Ma, X., & Yu, Z. (2019). Three dimensional characterization of meteorological and hydrological droughts and their probabilistic links. *Journal of Hydrology, 578*, 124016.

CHAPTER

6

Multivariate return period and quantile

In this chapter, we present the last step of the standard multivariate hydrological frequency analysis (HFA). It deals with the inference, in particular risk assessment in terms of return periods (RPs) or quantiles. Performing this step is based on the analysis and decisions from previous steps (described in previous chapters), especially the modeling step (Chapter 5). Here, we briefly present risk assessment in hydrology, followed by the basics regarding multivariate RP and quantile, and, finally, we present an overview of the statistical approaches and methods regarding the selection of the multivariate combinations and events for a given RP with illustrative examples.

6.1 Risk assessment in hydrology

Risk assessment is one of the most important objectives of hydrological studies for many engineering projects, such as dam and bridge design and construction. In this regard, RP is a simple but fundamental concept. RP is widely used in engineering practices and well known in water resources and civil engineering, as well as in other disciplines such as seismology, oceanography, geology, geophysical, and environmental sciences. Commonly, RP is defined as the average of the time intervals elapsing between two successive events, such as exceedances of a given threshold of river discharge. The concept of *design quantile* is also important and closely related to RP. It is usually defined as the value(s) of the variable(s) describing the event that is (are) associated with a given RP. In other words, the RP is inversely linked to the probability of exceedance of a specific value of the variable under consideration (e.g., flood peak), which, in turn, is related to the notion of quantile. Formally, we are interested in studying the probability $\Pr(X \geq x_T)$ of an event x_T to be exceeded

DOI: https://doi.org/10.1016/B978-0-323-95908-7.00005-0

for a random variable X from a distribution F, where x_T corresponds to the quantile of order p related to a RP T, such that

$$T = 1/(1 - \Pr(X \leq x_T)) = 1/(1 - F(x_T)) \text{ and } x_T = F^{-1}(1 - 1/T) \qquad (6.1)$$

The importance of the hydraulic structure and the possible consequences of its failure are among the aspects in engineering practice considered while chosing the RP value T. For instance, for a dam design, T is usually larger than 1000 years, whereas for a sewer construction, T can range between 5 and 10 years. Note that a T-year RP does not mean that a T-year event should happen every T years, but it shows the probability of the occurrence of a T-year event (e.g., flood) is 1/T every year.

In the univariate case, the use of the RP is simple, and it is based on (6.1). More precisely, first, a regulation RP T^* is identified; then, the associated critical probability level p^* is evaluated; and, finally, the selected distribution function F is inverted, leading to the critical design quantile $x^* = F^{-1}(p^*)$. The quantile x^* can be seen as a critical threshold in the sense that occurrences smaller than x^* are considered safe, whereas those larger than x^* are identified as dangerous. A regulation RP for dam design is usually fixed in some countries via national laws and guidelines. For instance, in Austria, RP is 5000 years, in France, depending on the dam typology, RP is between 1000 and 10,000 years, whereas in Spain, it ranges from 500 to 10,000 years (e.g., Requena et al., 2013). However, these regulations do not specify which variable (s) such as the peak, volume, or the duration of the flood is involved.

As mentioned above, risk assessment is considered as the last step in HFA. It requires to perform all the previous steps presented in previous chapters. Indeed, whatever be the adopted approach to evaluate the risk, it is required to select the most appropriate distribution (univariate or multivariate, Chapter 5), which, in turn, depends on basic assumption check (Chapter 4) as well as the preliminary analysis (Chapter 3). These specific concepts and steps are valid in a multivariate setting with appropriate adaptations. One of the most important and challenging aspects in the multivariate risk assessment is data set lengths, and making an appropriate selection among multiple options (e.g., variables to be involved, events of interest, RP, and quantile definitions).

6.2 Multivariate return periods and multivariate quantile: generalities

Unlike the univariate case, the notions of multivariate RP and quantile are not straightforward and faced with a number of challenges. Over the last decade, several efforts have been made to study issues related to multivariate design and quantiles. For a relatively recent review, see Gräler et al. (2013) and Serinaldi (2015).

As mentioned in previous chapters, when the random variables describing the event are correlated/dependent and are all relevant to risk evaluation or design purposes, it is recommended to consider the joint study of their probabilistic characteristics. This leads not only to a more accurate risk assessment compared to univariate analyses, but also to a better understanding of the phenomena under investigation. Moreover, if only the univariate RP is considered such as either flood peak or flood volume, then the risk related to a specific event can be over- or underestimated. In addition, the RP should be expressed in terms of risk of either dam overtopping or downstream damages, but not in terms of probability of occurrence of floods, in order to take into account the effect of reservoir and dam characteristics on the flood hydrograph routing process (Mediero et al., 2010). It is important to mention that while in the univariate case the RP of the unique variable involved equals the RP of structure failure, in the multivariate case, the latter is the result of interactions of combinations among the structure and the hydrological loads acting on it (Volpi & Fiori, 2014). In other words, the hydraulic structure may fail when a combination of the related hydrological variables exceeds a certain return level.

Based on a variety of statistical methods, different definitions of the multivariate RP are now available in the literature. Among them, some are based on regression analysis, bivariate joint distributions, bivariate conditional distributions, Kendall and survival Kendall distribution functions, and structure (described below in more detail). The majority of these definitions can be formulated through the copula function.

6.2.1 Definitions and presentation

In the following, we restrict our attention to the bivariate case. To this end, let X and Y be random variables such as the peak discharge and the flood volume, with joint distribution $F_{X,Y}$ given by $F_{X,Y}(x,y) = \Pr\{X \leq x \text{ AND } Y \leq y\}$, marginal distributions F_X and F_Y, respectively, and copula C. A general formulation of RP can be given by (e.g., Genest & Chebana, 2017; Salvadori et al., 2011)

$$T_A = \frac{\mu}{p_A} = \frac{\mu}{\Pr\{A\}} \tag{6.2}$$

where A is the critical (dangerous) event and μ is the average interarrival time between two realizations of the process. This formulation is general and flexible where the event A can take a variety of particular cases. In the literature dealing with the applications of HFA, one of the challenges is to adapt (6.2) to a multivariate setting. In the univariate case, we have $A = \{X > x\}$ or $A = \{X < x\}$ depending on whether we are interested in floods or in droughts, and, accordingly, the denominator in (6.2) is uniquely defined as 1-$F(x)$ or $F(x)$, respectively. For a given RP T, the value of x is unique and corresponds to the quantile of X. However, in a multivariate context, a number of options for A are of interest in hydrology, such as the event $A = \{X > x \text{ AND } Y > y\}$ corresponds to a simultaneous (AND) exceedance event, as well as the OR exceedance $\{X > x \text{ OR } Y > y\}$, conditional $\{X < x | Y < y\}$, and AND non-exceedance $\{X < x \text{ AND } Y < y\}$ events. For instance, the event $\{X < x \text{ AND } Y < y\}$ could be of interest to study droughts, whereas the event $\{X > x \text{ AND } Y > y\}$ is important to study floods.

It is possible to express more explicitly the probability $\Pr\{A\}$ in (6.2) in terms of copula C as well as the margins $U = F_X(X)$ and $V = F_X(Y)$. This is useful since we can take advantage of the modeling step based on copula (Chapter 5). Indeed, we have the following formulations for which we can have the corresponding RP, such as T_{AND}, T_{OR}, and so on using (6.2):

$$
\begin{aligned}
p_{\text{AND}} &= \Pr\{U > u \text{ AND } V > v\} = 1 - u - v + C(u,v) \\
p_{\text{AND,NonExceedance}} &= \Pr\{U < u \text{ AND } V < v\} = C(u,v) \\
p_{\text{OR}} &= \Pr\{U > u \text{ OR } V > v\} = 1 - C(u,v) \\
p_{\text{COND1}} &= \Pr\{U > u | V > v\} = (1-u)(1 - u - v + C(u,v)) \\
p_{\text{COND2}} &= \Pr\{U > u | V \leq v\} = 1 - C(u,v)/u \\
p_{\text{COND3}} &= \Pr\{U > u | V = v\} = 1 - \partial C(u,v)/\partial u \\
p_K &= \Pr\{C(U,V) > t\} = 1 - K_C(t) \\
p_S &= \Pr\{g(U,V) > z\} = 1 - F_Z(z)
\end{aligned}
\tag{6.3}
$$

These expressions, except the last two ones, describe the different probabilities of exceedance events. However, the last equation presents p_S, the so-called "structure-based" RP introduced by Volpi and Fiori (2014), where g is a functional relationship linking the forcing (environmental) variables with the design (structural) variable $Z = g(U,V)$. The probability p_K corresponds to the so-called "secondary" or "Kendall" RP introduced by Salvadori and De Michele (2004). We can also find other methods, such as the one based on survival copula (Salvadori et al., 2013). Selecting the "best" RP definition appropriate for a design is still an open and debated question. Indeed, this choice would depend on how a given system (e.g., a hydraulic device) responds to a specific forcing.

As an explicit example, we consider the OR event corresponding to the probability of $\Pr\{X > x \text{ OR } Y > y\}$. We have

$$T_{OR} = \frac{\mu_T}{1-p} = \frac{\mu_T}{1 - F_{X,Y}(x_p, y_p)} = \frac{\mu_T}{1 - C_{UV}(u_p, v_p)} \tag{6.4}$$

for a given critical probability level p, where x_p, y_p are the combinations of any point belonging to the critical level curve L_p as the set of points:

$$L_p = \{(x, y) \in R^2,\ F_{X,Y}(x, y) = p\} \tag{6.5}$$

as illustrated in Fig. 6.1.

In multivariate HFA, the focus was paid on the multivariate RP. However, in HFA, the RP and quantile are closely related. In the statistical literature, the notion of multivariate quantile is not new. The univariate quantile is generalized to the multivariate setting in a number of ways. However, the interpretation of the obtained quantile values is one of the main challenges of multivariate quantile extensions. Chebana and Ouarda (2011) studied the notion of multivariate quantiles in hydrology. The selected multivariate quantile version has some advantages (simple, intuitive, interpretable, and probability-based). In addition, this multivariate quantile version is related to the copula function as well as to the level curve L_p in (6.5) employed in the context of multivariate RP. Properly, this bivariate quantile version is a curve composed by combinations (x,y) that fulfil $F_{X,Y}(x, y) = p$ (an

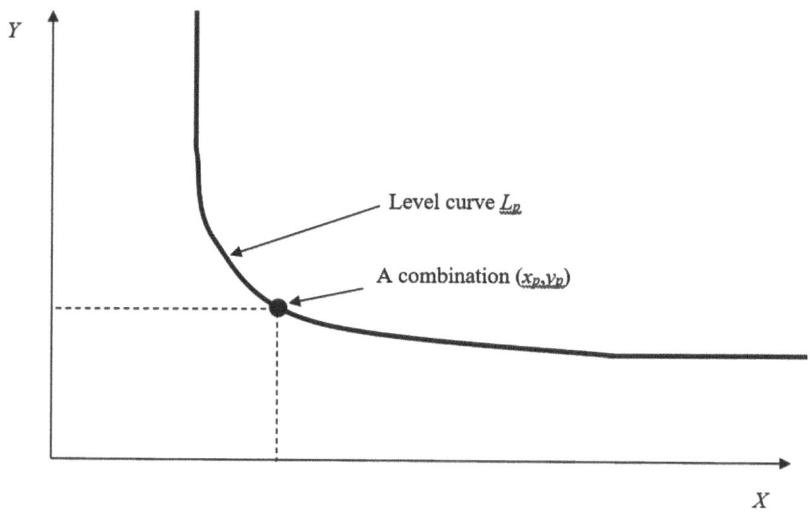

FIGURE 6.1 Illustration for a given p of the level curve L_p and one of the infinite combinations (x_p, y_p) for the AND non-exceedence event $\{X < x \text{ AND } Y < y\}$.

infinity of combinations). Even though we present the bivariate case, all the elements of the developments for higher dimensions can be found in Chebana and Ouarda (2011).

Unlike the univariate case, where the quantile is a real value, the bivariate quantile is a curve. Indeed, from (6.4) and within the selected bivariate quantile version, infinite pairs (called also combinations) of values (x,y) share the same joint probability. For a given event A, all the bivariate combinations (x,y) have the same RP $T(p)$. However, these combinations are not necessarily equivalent from a practical or hydrological points of view. The hydrological design event is defined once a combination (x_p, y_p) is selected within the quantile curve, that is, an infinite set of events having the same joint probability value p. This can be seen as a flexibility for practitioners where they are able to select one or more convenient design combinations according to the needs of a specific application, taking into account "*the environment in which a structure should be designed, as well as the stochastic dynamics of the phenomenon under investigation*" (Salvadori et al., 2011). Moreover, several criteria can be employed for the selection of (x_p, y_p) such as the probability of occurrence, structure-based approach, regression or conditional probability, and routed RP. This is the objective of the methods presented in Section 6.3.

In light of the variety of possible options of multivariate RP definition, it is of interest to explore the possible relationships between the corresponding T values. One of the goals is to identify the most appropriate choice as well as to compare T values and associated return levels. However, it is shown that such studies basically do not have solid foundations and could be misleading. Indeed, beyond the RPs as numerical values, the above-defined multivariate RPs (e.g., T_{OR} and T_K) do not provide answers to the same problem statement. Therefore, the practical implications of the selected approach have to be carefully considered. One can find some comparisons between the different RPs (mainly T_{OR}, T_K, and T_{AND}), for instance, in Vandenberghe et al. (2011) and Gräler et al. (2013). Serinaldi (2015) provided an interesting discussion regarding comparisons of multivariate RPs.

The uncertainty of bivariate quantile estimation is crucial and also plays a prominent role in HFA. Indeed, uncertainty analysis is relevant to the reliability of both theoretical results and practical applications. To estimate this uncertainty, a number of algorithms and methods have been proposed, including a Bayesian method and a Copula-Based Parametric Bootstrapping Uncertainty (C-PBU) method. The source of major uncertainty in bivariate quantile estimation is sampling uncertainty due to data size limitation. On the other hand, uncertainties related to event selection and parameter estimation are of minor importance.

6.2.2 Illustrative example 1

This example is based on the dataset already employed in previous chapters. We used the obtained results from those chapters such as the descriptive analysis and the detected outlier of 1990 (Chapter 3), the basic assumptions (Chapter 4), and the selected copulas and margins (Chapter 5). Recall that from Chapter 5, two copulas were selected: Clayton copula when we do not focus on the tail dependence and the Husler–Reiss copula if the tail dependence is considered. When it comes to the margins, the detected outlier mainly has an impact on the distribution of the peak variable, which is a Gamma distribution by including the outlier and a Gumbel distribution by excluding the outlier. In addition, for comparison purposes, we consider two RPs $T = 10$ and 100 years.

Fig. 6.2 shows the level curves on the uniform space (0,1) for the same copula (Clayton) with different events (OR, COND1, and COND2) based on (6.3). We observe the difference in shape and the direction of the curves. However, Fig. 6.3 shows that the shape of the level curves

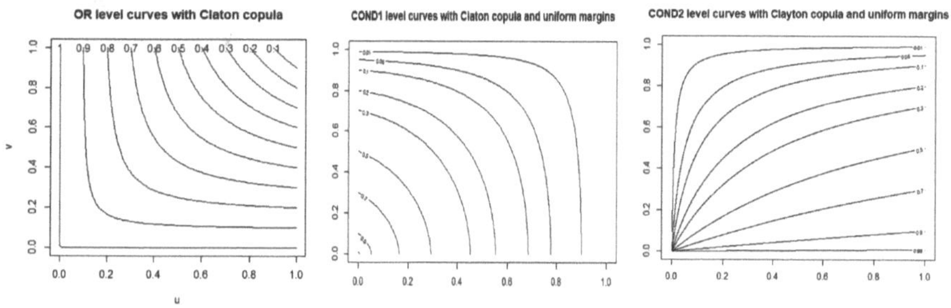

FIGURE 6.2 Levels for different events (OR, COND1, and COND2) for the Clayton copula on the uniform space.

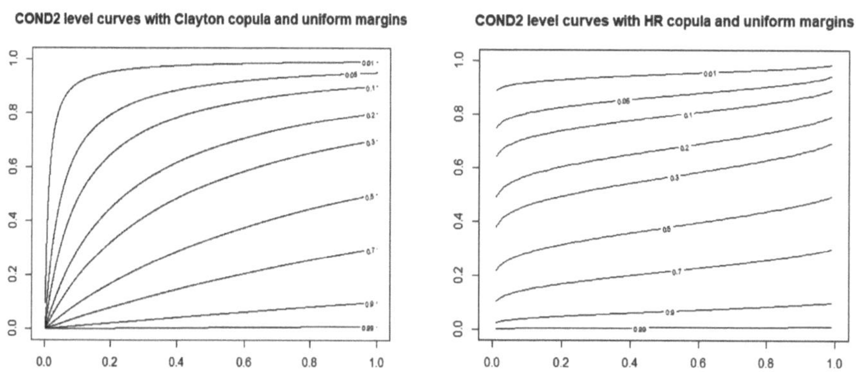

FIGURE 6.3 Level curves on the square space with Clayton and Husler–Reiss (HR) copulas corresponding to the event COND2 on the uniform space.

could change dramatically with the copula (here Clayton and Husler–Reiss copulas) given the same event (here the COND2 event).

In Fig. 6.4, we present the level curves on the space of the original variables flood peak Q and volume V as in (6.5). We also present the univariate quantiles of Q and V for each RP value.

In Fig. 6.5, we present the level curves for both copulas Clayton and Hussler-Reiss (Fig. 6.5A and B respectively for $T = 10$ and 100 years) and two different events OR and K (Fig. 6.5C for $T = 10$ years) including and excluding the detected outlier in Chapter 2 (year 1990). We observe that the outlier has negligible impact on the level curves as well as on the univariate quantile values except for the peak where the impact is significant with high RP (here $\underline{T} = 100$). We also observe another aspect on the shape of the curves, which is significantly different from one copula to another for the same RP and same event (here with OR event and from Clayton and Husler–Reiss). Finally, for a given copula, the shape of the level curve varies from one event to another (here OR and K, Fig. 6.5C). These observations indicate the importance on the risk assessment of selecting the appropriate copula as well as choosing the event of interest for the project at hand.

Fig. 6.6 more explicitly shows these elements by focusing on the same RP ($T = 100$ years) but with other events (AND, AND nonexceedance, COND1 and COND2). According to the event, the level curves are somehow distant especially for the AND nonexceedance but not uniformly in some parts. The univariate quantiles are reported on the AND nonexceedance event (Fig. 6.5B) as the circles on the axes.

6.3 Methods to select combinations

As described in the previous section, in multivariate HFA, an appropriate RP for structure design leads to an infinite number of critical combinations of the related random variables. Even though these combinations are

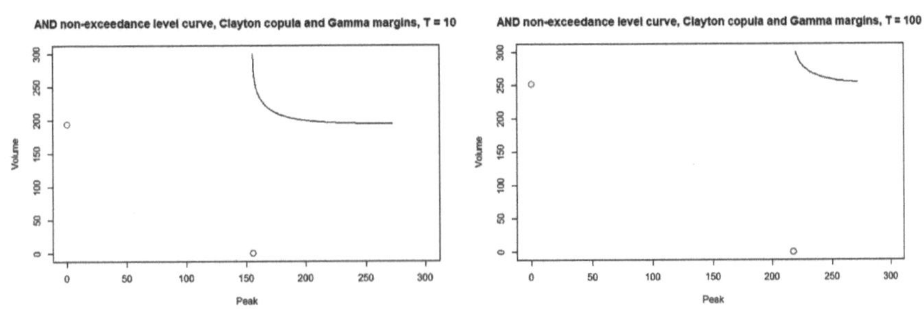

FIGURE 6.4 AND nonexceedance level curves associated with $T = 10$ and 100 years with Clayton on the space of the original variables Q and V. The small circles on the axes are the corresponding univariate quantiles.

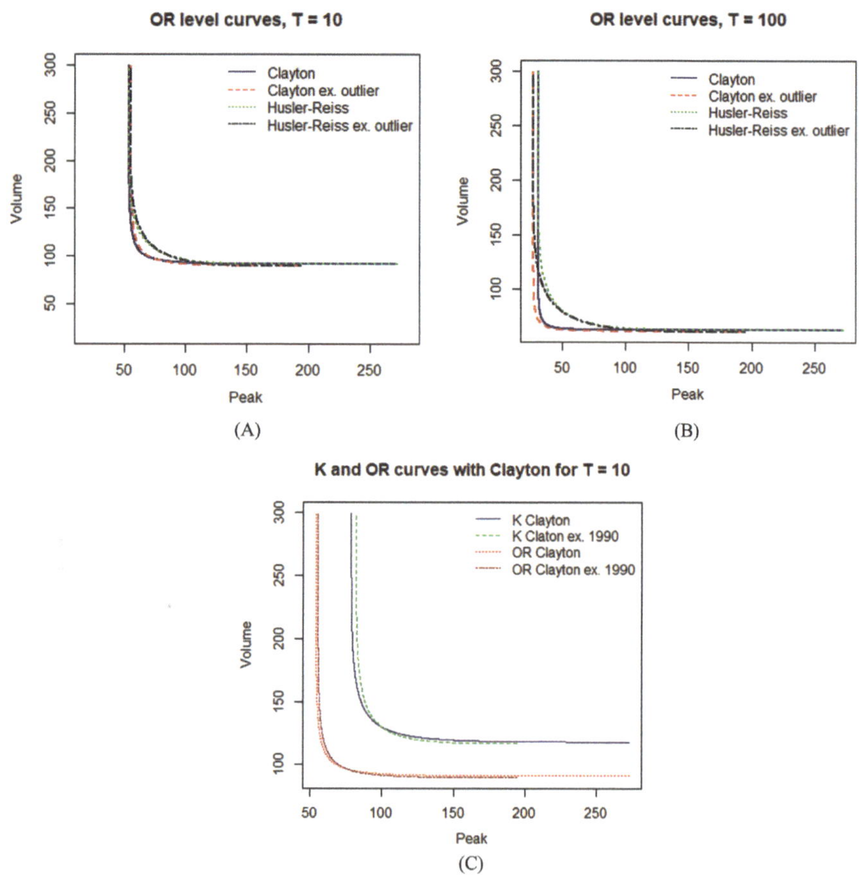

FIGURE 6.5 Level curves on the space of the original variables Q and V with and without the outlier, associated with $T = 10$ and 100 years with Clayton (A) and Husler–Reiss (B), and both Kendall and OR events for Clayton (C).

statistically equivalent (leading to the same risk), they are generally not equivalent in practice and are dependent on the hydrological applications. According to the project as well as the available resources, one or more design combinations could be designated to evaluate the effects of different hydrological loads on a structure. This provides flexibility to designers and engineers over the univariate framework in addition to the multivariate framework being more accurate and realistic (as discussed in previous chapters).

For a design event, characterized by several variables, and for a given value T of the RP, methods to select a section of the level curve or a specific combination (x_p, y_p) are needed. These methods can be classified in two main categories. In the first one, a single combination (x_p, y_p) is

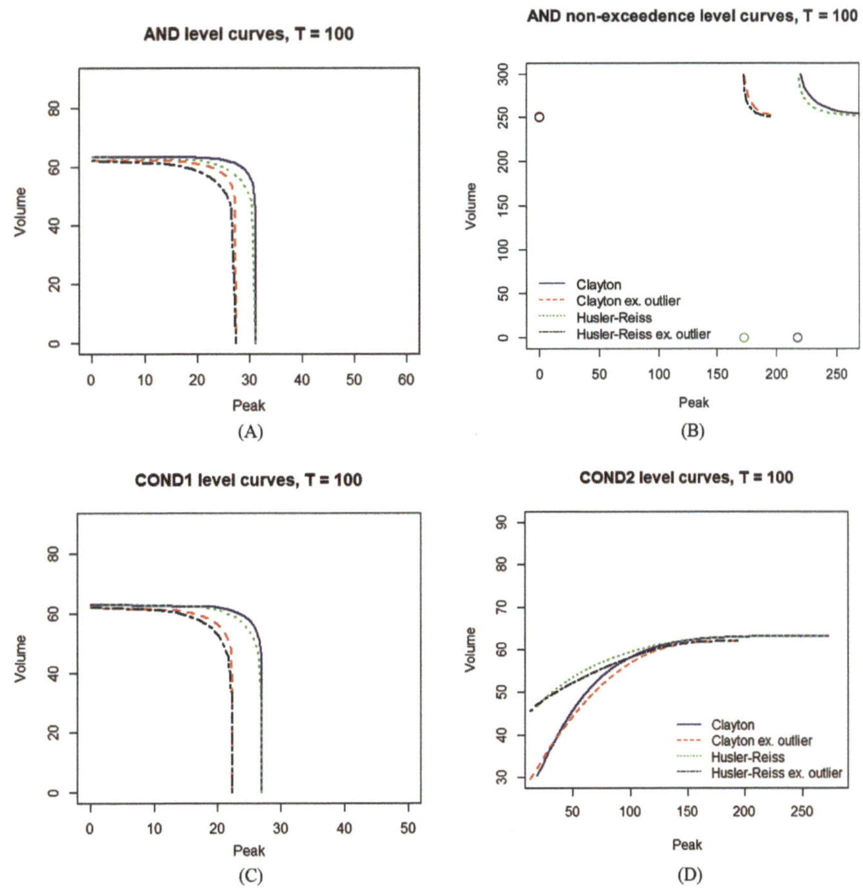

FIGURE 6.6 Level curves for $T = 100$ and different events: AND (A), AND nonexceedance (B), COND1 (C), and COND2 (D) associated with Clayton and Husler–Reiss copulas. The circles on (B) represent the univariate quantiles for the peak and volume (including and excluding the outlier).

selected. In this category, the multivariate RP or quantile is transferred to the univariate setting either through a real-value transformation (e.g., structure-based, Kendall RP), conditioning over one variable (e.g., regression and conditional approaches), or with an additional criterion (e.g., most likely design realization approach). In the second category, the aim is to make a subset selection of combinations (ensemble-based) from the quantile (level) curve (e.g., quantile proper part and alpha approaches). Even though useful for practitioners, the selection of a unique single combination reduces the amount of information that can be obtained by the multivariate framework. However, the importance

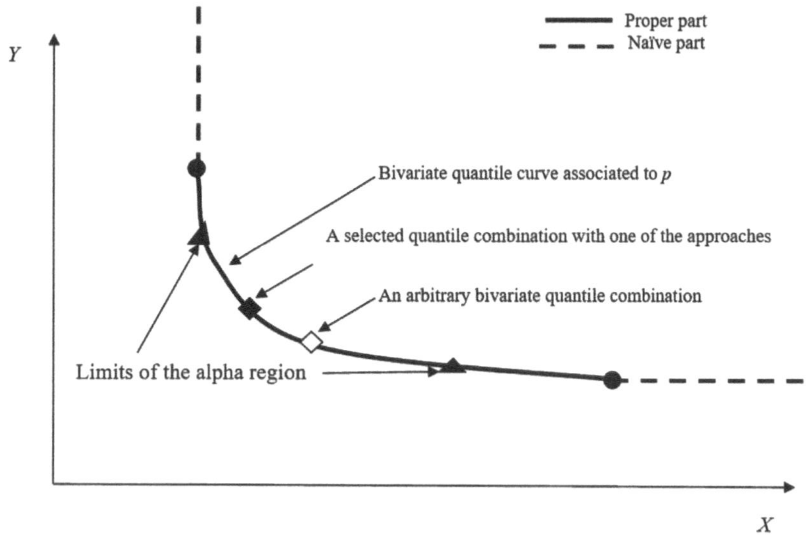

FIGURE 6.7 Illustration of the main elements of the approaches to select combinations.

and richness of the ensemble-based approaches is highlighted in a number of studies. Fig. 6.7 provides a generic illustration of the main elements of those approaches. In the following, we present a number of these methods and then some of them will be illustrated in an example in the last subsection.

6.3.1 Most likely design realization approach

This approach aims to select a single design combination (x_p,y_p), which maximizes a weight function (Salvadori et al., 2011). This function attributes weights to the realizations lying on the critical level curve, of which two examples of weight functions are proposed by Salvadori et al. (2011). In the first one, the critical combination is characterized by the highest value of the joint probability of exceedance. It is called *"component-wise excess design realization"* and should be seen as statistically *"safety lower-bound"*. In the second weight function, the obtained combination is the *"most-likely design realization"* and characterized by the highest value of the joint bivariate probability density function f_{XY}. The latter, is maximized over the t level curve in order to select the most likely point. Usually, not all pairs or combinations (u,v) on the t level curve have the same likelihood where pairs on the edges are less likely than those closer to the center of the level curve.

More explicitly, consider, for instance, the T_{OR} RP where all combinations (u,v) are located at the same probability level $t_{OR} = C_{UV}(u,v)$ of the copula, then they have the same bivariate RP T_{OR}. For a given design RP with the corresponding level t_{OR}, the most likely design combination (u_{OR}, v_{OR}) of all possible ones at this level can be obtained by selecting the combination with the highest joint probability density:

$$(u_{OR},\ v_{OR}) = \underset{C_{UV}(u,v)=t_{OR}}{\operatorname{argmax}}\ f_{XY}\left(F_X^{-1}(u),\ F_Y^{-1}(v)\right) \quad (6.6)$$

Then, the corresponding combination is composed of design (quantile) values x_{OR} and y_{OR}, which are $(x_{OR},\ y_{OR}) = \left(F_X^{-1}(u_{OR}),\ F_Y^{-1}(v_{OR})\right)$. In general, provided that weak regularity conditions are satisfied, the function $f_{X,Y}$ can be calculated by using the marginal densities f_X and f_Y as well as the copula density function $c(.,.)$

$$f_{X,Y}(x,y) = c\left(F_X(x), F_Y(y)\right)f_X(x)f_Y(y) \quad (6.7)$$

Other events, such as the AND event, can be also considered. Due to the complexity of deriving analytical solutions in (6.6), numerical methods such as Newton's harmonic mean can be considered to estimate the most likely realizations method (see Yin et al., 2018).

6.3.2 Structure-based approach

Inspired by the univariate setting, Volpi and Fiori (2014) proposed a structure-based approach. Indeed, in the univariate setting, the structure design variable Z is related to the hydrological load X as $Z = g(X)$ through a strictly increasing function $g(.)$ denoted as the structure function. In the univariate case, both RPs based on design variable Z and hydrological load X are identical. However, in the bivariate setting, the design variable Z depends on both hydrological variables X and Y, such that $Z = g(X,Y)$. Here, the function $g(.,.)$ takes into account the interactions among the structure and the hydrological loads acting on it. The design variable Z may represent directly the structure dimension, such as the size of a spillway or the elevation of a levee, or another quantity related to the structure. Unlike the univariate case, the multivariate RP of structure failure depends on the particular structure under design.

In practice, the function g measures the effects of the hydrological load on the structure. However, more complex formulations can be considered. In their illustrative example, Volpi and Fiori (2014) provided a function g given by

$$g(Q,V) = Q\left[1 - e^{-SQ/V}\right] \quad (6.8)$$

for the flood variables peak Q and volume V with a given constant S, which depends on the structure under study. The probability distribution function of Z is obtained as

$$F_Z(z) = \int_{D_Z} f_{X,Y}(x,y)dxdy \tag{6.9}$$

where D_z is the region of the pairs (x,y) such that $D_z = \{(x,y) \in R^2, g(x,y) \leq z\}$.

Since Z is a real-valued variable, the RP of structure failure can be obtained as in the univariate HFA by writing $T(z) = \mu/(1 - F_Z(z))$. However, this RP is the mean time elapsed between two successive events belonging to the supercritical bivariate region $\overline{D}_z = \{(x,y) \in R^2, g(x,y) > z\}$. Therefore, given an appropriate value T of the RP of structure failure, the corresponding design quantile is $z_T = F_Z^{-1}(1 - \mu/T)$.

As indicated by Volpi and Fiori (2014), this method can be generalized to the case where Z is a vector of design variables. The corresponding region D_z could be not connected. Since this method can be seen as a dimension reduction technique, its use can be unsuccessful if the design (structural) variable is not unique.

6.3.3 Kendall return period approach

Salvadori and De Michele (2004) proposed what they called the Kendall or the secondary RP. It shares some of the motivations behind the structure-based approach. Indeed, the idea behind this concept is to transfer the multivariate RP to a univariate one through the use of the Kendall function $K(.)$. It measures the probability of an event to occur in one of the two unique sub-regions given through a level curve with a unique value of joint probability. In this context, the structure-based RP allows us to further expand the above discussion. Note that the structure-based and Kendall RPs have the same form where the latter can be seen as a particular case of the former with a specific choice of g. Hence, as it is indicated for the structure-based approach, the Kendall RP, seen also as a dimension reduction technique, can be unsuccessful if the design (structural) variable is not unique. Salvadori et al. (2011) extended the bivariate RP to the multivariate setting.

From the general formulation in (6.2), the Kendall RP is given by

$$T_K = \frac{\mu_T}{1 - K(t)} \tag{6.10}$$

where K is Kendall's distribution function $K(t) = \Pr\{F_{XY}(x,y) \leq t\}$ associated with the joint distribution function $F_{X,Y}$. As in the structural-based approach, this function is a univariate representation of multivariate information, and this RP corresponds to the mean interarrival time of the supercritical events, which include all the events belonging to the bivariate region $D_t = \{(x,y) \in R^2, F_{X,Y}(x,y) > t\}$.

Given T_K the design RP, the corresponding probability level t_K can be obtained as in the univariate case:

$$t_K = K^{-1}\left(1 - \frac{\mu_T}{T_K}\right) \tag{6.11}$$

Once t_K is obtained, the most likely design combination (u_K, v_K) in $[0, 1]^2$ is selected on the associated level curve, similarly as for the OR event described by (6.4). Using the inverse of the marginal distributions F_X and F_Y, the corresponding design event (x_{KEN}, y_{KEN}) is found.

In a number of situations where no analytical expression for the function K is available, its inverse can be obtained numerically. In addition, this RP is more suitable for Archimedean copulas but not appropriate for extreme value copulas since the latter have all the same function K, as mentioned in Chapter 5.

6.3.4 Conditional distribution approach

This can be seen as a natural approach in case we have a specific value of one of the variables, such as a quantile of particular interest related to a univariate RP. It consists of conditioning the bivariate distribution function $F_{XY}(x,y)$ on the univariate marginal design quantile $x_{MAR} = x_{UNI}$ corresponding to T_{MAR} of the selected univariate design RP (Gräler et al., 2013). The obtained univariate conditional distribution function $F_{Y|X}(y \mid x = x_{MAR})$ can be employed to get the value y_{MAR} for the conditional univariate design RP T_{MAR}. As indicated above, the formulation of this RP can be expressed in terms of copula $C_{UV}(u,v)$. Indeed, with $u_{MAR} = F_X(x_{MAR})$ and $v_{MAR} = F_Y(y_{MAR})$, we have

$$T_{MAR} = \frac{\mu_T}{1 - F_{Y|X}(y|x = x_{MAR})} = \frac{\mu_T}{1 - C_{V|U=u_{MAR}(v_{MAR})}} \tag{6.12}$$

where $U = F_X(X)$ and $V = F_Y(Y)$. The corresponding level is

$$v_{MAR} = C^{-1}_{V|U=u_{MAR}}\left(1 - \frac{\mu_T}{T_{MAR}}\right) \text{ and finally } y_{MAR} = F_Y^{-1}(v_{MAR}) \tag{6.13}$$

As it is defined, this approach cannot be considered as bivariate joint RP in the strict sense.

6.3.5 Regression-based approach

This approach is motivated by situations where a dominant driving variable can be selected in the practical application (Gräler et al., 2013). In this approach, first, the driving variable X is chosen, which represents the prevailing variable in the design. Then, given T_{REG} a design RP, using the marginal distribution F_X of X, the corresponding design quantile x_{REG} is obtained from

$$T_{\mathrm{REG}} = \frac{\mu_T}{1 - F_X(x_{\mathrm{REG}})} \tag{6.14}$$

In a regression modeling context with X as an explicative variable over the other design variable of interest Y, the corresponding design value y_{REG} is obtained using a regression function f_{REG} as $y_{\mathrm{REG}} = f_{\mathrm{REG}}(x_{\mathrm{REG}})$. The regression function f_{REG} can be linear. Note that this approach does not correspond to a joint RP in the proper sense, but rather it is based on a univariate frequency analysis. It has been applied, for example, by Serinaldi and Grimaldi (2011).

It is important to mention the similarity between the regression and the conditional approaches. Indeed, the regression-based method provides a predicted value for Y given a certain quantile of X, whereas the conditional method estimates the quantile of Y conditioned under the quantile of X. Hence, from a probabilistic point of view, both approaches cannot directly be compared.

6.3.6 Multivariate quantile curve, proper part approach

As mentioned above, a number of multivariate quantile definitions are available in the statistical literature. Among them, the adapted version to HFA applications is simple, intuitive, probability-based, interpretable, can be expressed with copula, and coherent with the multivariate RP notion in HFA. It is a curve expressed as follows for p in (0,1) and when considering the *AND non-exceedence* event $\{X \le x \text{ AND } Y \le y\}$:

$$Q_{X,Y}(p) = \left\{(x,y) \in R^2 \text{ such that } x = F_X{}^{-1}(u), y = F_Y{}^{-1}(v); u, v \in [0,1]: C(u,v) = p\right\} \tag{6.15}$$

Other events, as presented earlier, can also be considered such as $\{X \ge x \text{ AND } Y \ge y\}$ and $\{X \ge x \text{ OR } Y \ge y\}$. For those events, formula (6.15) can be adapted based on the expressions in (6.3) using copula.

In the context of HFA, this approach consists of the decomposition of the quantile curve into a *naive* part in the tail of the curve and a *proper* part as the center of the curve. The former is composed of two pieces beginning at the end of each boundary of the proper part. In practical applications, it is important to specify the points that define the extremities (asymptotes) of the proper part. In this senses, note that the common univariate quantiles are particular cases of the bivariate quantile curve representing the extreme points (asymptotes) of the proper part of the bivariate quantile. In addition, the univariate estimated quantiles, correctly combined, are particular cases corresponding to the extreme scenarios of the bivariate quantile curve. For a given bivariate sample, univariate quantile values should be used carefully since their direct combination generally does fall in the curve and hence does not represent the desired level of risk. This may lead to inappropriate conclusions.

Note that the terms of level or isoline are employed in a number of multivariate HFA studies without specifically talking about bivariate quantile curve. The bivariate case presented here is also available in higher dimensions (see Chebana & Ouarda, 2011).

6.3.7 Alpha-region approach

This methodology, proposed by Volpi and Fiori (2012), aims to select a subset from the critical level curve. This subset consists of pairs (combinations) based on their probability of occurrence. More precisely, given a percentage quantity α, it is divided into the levels of probability α_1 and α_2 ($\alpha = \alpha_1 + \alpha_2$) that define, respectively, the upper and lower bounds of the subset. Their values can be suggested according to the specific application, such as eliminating events with the lowest probability of occurrence. It can also be seen that this subset corresponds to a section of the proper part of the level curve (as in the bivariate quantile curve). According to Volpi and Fiori (2012), hydraulic studies have shown that the critical combinations for the design of flood-control reservoirs, such as in their case study of the Tiber River, were close to the lower pair in the curve (with $T = 200$ years and $\alpha = 0.05$ in their application).

In practice, once the subset is defined, several combinations can be chosen within the subset, resulting in an ensemble approach. These combinations can be employed to assess the effects of different hydrological loads on the structure. Among this ensemble of combinations, one can choose, among others, the combination that is most critical for the structure. In their application, Volpi and Fiori (2012), dealing with the Tiber River, the lower bound combination was chosen as the design event. Indeed, hydraulic studies highlighted that this combination is the most critical for the design of the flood control reservoirs.

6.3.8 Ilustrative example 2

We continue with the dataset from previous chapters (in particular Chapters 2 and 5) and previous section as well. Given the data and analysis in Chapter 5, recall that the selected multivariate distribution is composed of a copula (either Clayton or Husler–Reiss) and margins (either Gamma or Weibull).

To go further based on the previous quantile curves and RPs (obtained in Illustrative example 1), in the following, we focus on combination selection. First, we treat the most likely realization design. To this end, we start by obtaining the joint density in (6.7). Since the latter is a surface, at given levels (RPs $T = 10$ and 100 years), Fig. 6.8 presents a number of such densities associated with different copulas (Clayton and Husler–Reiss). We observe that the density shape became sharper if T is large (comparing $T = 10$ and 100). In addition, the density f_{xy} is different from Clayton and Husler–Reiss copulas where, in particular, the one from Clayton is sharper than the one from Husler–Reiss for the same T.

Based on the density in (6.7) plotted in Fig. 6.8, we obtained the most likely combination for a number of situations. The latter are plotted in Fig. 6.9 where the circles on the OR curve are the most likely design realizations (combinations) for each case.

The other option to select a single combination from the quantile curve is based on the Kendall RP and level given in (6.11). Fig. 6.10

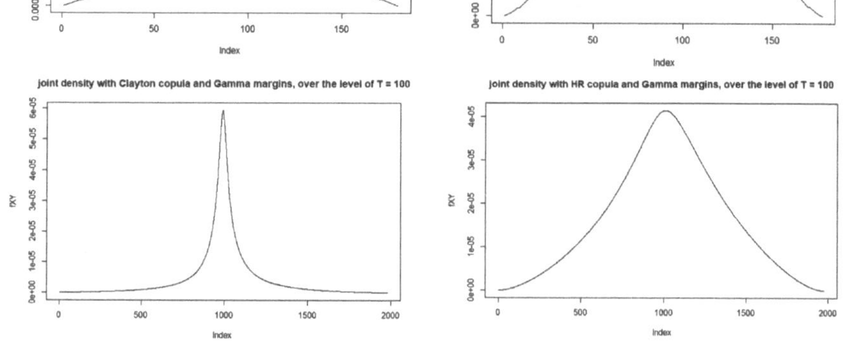

FIGURE 6.8 Levels of the joint density f_{xy} given in (6.7) for $T = 10$ and 100 years with Clayton and Husler–Reiss copulas.

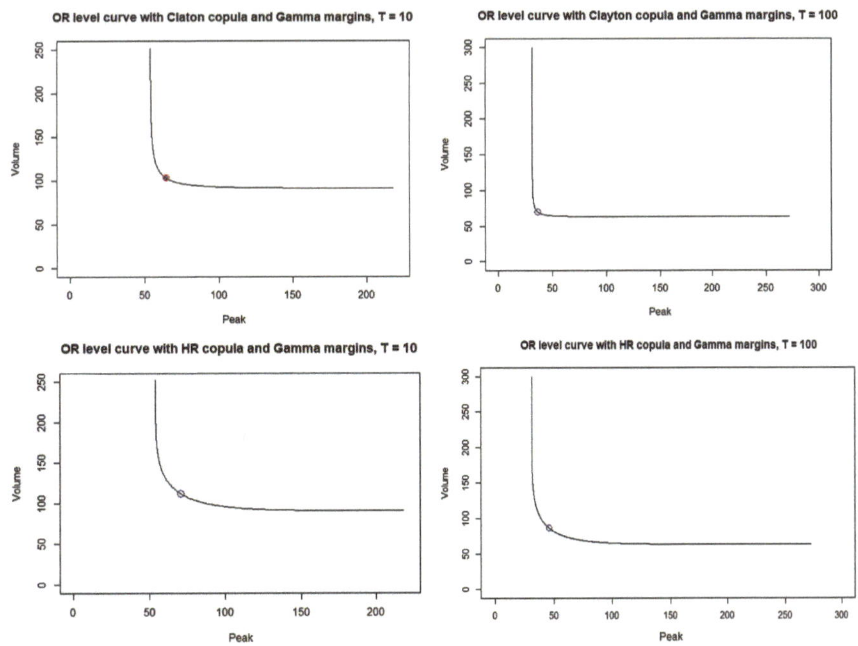

FIGURE 6.9 The most likely realizations (combinations) on their corresponding quantile curves for Clayton and Husler–Reiss copulas, $T = 10$ and 100 years.

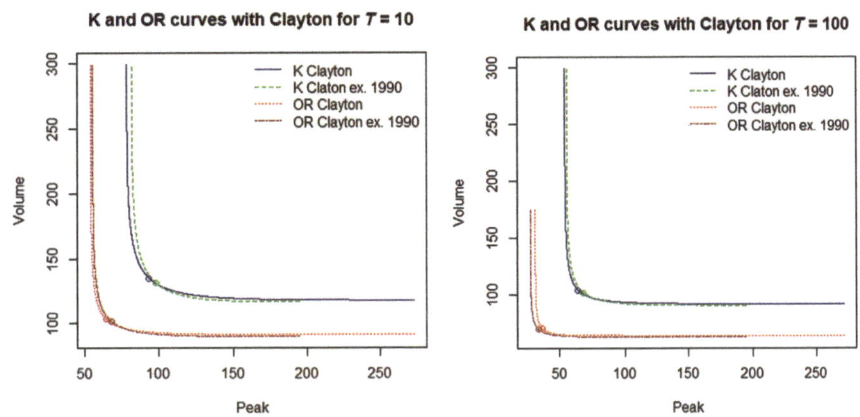

FIGURE 6.10 Quantile curves for the Clayton copula with the Kendall return period and the OR event along with including or excluding the detected outlier (left $T = 10$ and right $T = 100$ years).

presents the obtained curves for Kendall and, for comparison purposes, also those with the OR event. Note that the Kendall curves, and hence the associated combinations, are not available for the extreme value

TABLE 6.1 Univariate and multivariate realizations with different multivariate approaches.

		Univariate		Multivariate					
T		Q	V	**Copula**	**COND1**	**COND2**	**COND3**	**Kendall**	**Most likely (OR)**
10	All data	155.10	194.15	Clayton	(39.88, 123.11)	(52.73, 150.41)	(49.21, 143.74)	(93.49, 135.18)	(64.53, 103.42)
				Husler–Reiss	NA	NA	NA	NA	(70.51, 112.51)
	Excluding 1990	140.05	192.93	Clayton	(44.89, 134.75)	(26.54, 98.02)	(24.60, 93.76)	(97.82, 131.74)	(68.08, 101.54)
				Husler–Reiss	NA	NA	NA	NA	(77.14, 107.47)
100	All data	217.85	251.56	Clayton	(24.27, 125.24)	(31.12, 155.40)	(37.42, 177.40)	(64.26, 103.86)	(36.20, 69.89)
				Husler–Reiss	NA	NA	NA	NA	(45.76, 86.22)
	Exc. 1990	172.46	250.57	Clayton	(19.04, 125.29)	(27.40, 155.46)	(26.76, 153.44)	(68.23, 101.64)	(33.79, 68.76)
				Husler–Reiss	NA	NA	NA	NA	(47.75, 82.47)

NA: not available for all copulas because of the conditional copula and its inverse formulas in (6.12) and (6.13) are not available. Currently, these formulas are available for Archimedean and elliptical copulas.

copulas (here Husler–Reiss). This is why Fig. 6.10 is only for the Clayton copula (and not the Husler–Reiss one). Even though in principle it is the same T, the curves in Fig. 6.9 are shifted based on Kendall or the OR events. The reason is that the Kendall event is associated with a t_K as a quantile order (which, in turn, is associated with an RP), which is different from p. Indeed, in the case of $T =$ 10 and 100 years, we have $p =$ 0.90 and 0.99, whereas respectively t_K = 0.683 and 0.899, based on (6.11).

Table 6.1 summarizes the values of the combinations for different approaches and situations. First, note that in the multivariate case, we have a variety of options associated with the same T, whereas in the univariate case, we have only one combination (which can be seen as associated with the independence copula). We also observe that the effect of the outlier appears for $T = 100$, which confirms previous results. For the conditional approaches (COND1, COND2, and COND3), we need to provide a value of the first variable, here, for instance, we take the peak Q, and then based on (6.13), we obtain the associated value of the second variable, here the volume V. In Table 6.1, we provide arbitrary values for Q and hence obtain the associated value of V. However, note that the formulas for these equations are not available for the Husler–Reiss copula.

As expected and observed in previous results, it confirms here that the effect of the outlier is not high. However, the effect of the tail dependence is important and become even more with larger T.

References

Chebana, F., & Ouarda, T. B. (2011). Multivariate quantiles in hydrological frequency analysis. *Environmetrics (London, Ont.), 22*(1), 63–78.

Genest C. and Chebana F. (2017). Chapter 30: Coupula modeling in hydrologic frequency analysis. Handbook of Applied Hydrology, 2nd ed.; Singh, V.P., Ed.

Gräler, B., Van Den Berg, M., Vandenberghe, S., Petroselli, A., Grimaldi, S., Baets, B. D., & Verhoest, N. (2013). Multivariate return periods in hydrology: A critical and practical review focusing on synthetic design hydrograph estimation. *Hydrology and Earth System Sciences, 17*(4), 1281–1296.

Mediero, L., Jiménez-Álvarez, A., & Garrote, L. (2010). Design flood hydrographs from the relationship between flood peak and volume. *Hydrology and Earth System Sciences, 14* (12), 2495–2505.

Requena, A., Mediero, L., & Garrote, L. (2013). A bivariate return period based on copulas for hydrologic dam design: Accounting for reservoir routing in risk estimation. *Hydrology and Earth System Sciences, 17*(8), 3023–3038.

Salvadori, G., & De Michele, C. (2004). Frequency analysis via copulas: Theoretical aspects and applications to hydrological events. *Water Resources Research, 40*(12).

Salvadori, G., Durante, F., & De Michele, C. (2013). Multivariate return period calculation via survival functions. *Water Resources Research, 49*(4), 2308–2311.

Salvadori, G., Michele, C. D., & Durante, F. (2011). On the return period and design in a multivariate framework. *Hydrology and Earth System Sciences, 15*(11), 3293–3305.

Serinaldi F. (2015). Dismissing return periods! Stochastic environmental research and risk assessment 29(4), 1179–1189.

Serinaldi, F., & Grimaldi, S. (2011). Synthetic design hydrographs based on distribution functions with finite support. *Journal of Hydrologic Engineering, 16*(5), 434–446.

Vandenberghe, S., Verhoest, N., Onof, C., & De Baets, B. (2011). A comparative copula-based bivariate frequency analysis of observed and simulated storm events: A case study on Bartlett-Lewis modeled rainfall. *Water Resources Research, 47*(7).

Volpi, E., & Fiori, A. (2012). Design event selection in bivariate hydrological frequency analysis. *Hydrological Sciences Journal, 57*(8), 1506–1515.

Volpi, E., & Fiori, A. (2014). Hydraulic structures subject to bivariate hydrological loads: Return period, design, and risk assessment. *Water Resources Research, 50*(2), 885–897.

Yin, J., Guo, S., He, S., Guo, J., Hong, X., & Liu, Z. (2018). A copula-based analysis of projected climate changes to bivariate flood quantiles. *Journal of hydrology, 566*, 23–42.

CHAPTER

7

Multivariate nonstationary frequency analysis

7.1 Nonstationary hydrological frequency analysis

The traditional hydrological frequency analysis (HFA) assumes that the data series are stationary, among other assumptions (more details can be found in Chapter 4). A data series can be considered stationary if over time it does not show any significant changes. The statistical characteristics of hydrological data series might be altered under nonstationarity. Hence, because of a number of different reasons, it is not reasonable to always assume the stationarity is fulfilled. Indeed, climate change, urbanization, and deforestation are some of those reasons. Based on theoretical considerations and observational evidences, it is no longer valid to assume a flood design is stationary, which has been confirmed by the recent literature (e.g., Bracken et al., 2018; François et al., 2019; Salas et al. (2018)). As a result, it is essential to first test the validity of stationarity of the data series; otherwise, the models should be adapted to account for nonstationarity conditions. To this end, nonstationary models are introduced and developed. These models allow us to incorporate nonstationarity concerns into risk assessment frameworks. If nonstationarity is not taken into account, the results of the traditional HFA would be invalid. In that case, tasks of water management should be adapted accordingly. Hence, under ever-changing conditions, nonstationary HFA is becoming indispensable for hydrology design.

As presented in previous chapters, a large part of hydrological phenomena are described by two or more correlated/dependent characteristics (see e.g., Chapters 2 and 5 for more details), such as peak, duration, and volume of floods. The dependence between these features is important, which is considered in a multivariate HFA framework.

Multivariate Frequency Analysis of Hydro-Meteorological Variables
DOI: https://doi.org/10.1016/B978-0-323-95908-7.00007-4

TABLE 7.1 Selected references for the different frequency analysis frameworks.

	Univariate	Multivariate
Stationary	Rao and Hamed (2000)	Chapters 5 and 6 and references therein
Nonstationary	Strupczewski et al. (2001), Wang et al. (2014), El Adlouni et al. (2007), Gilroy and McCuen (2012)	Bender et al. (2014), Jiang et al. (2015), Kwon and Lall (2016), Sarhadi et al. (2016), Zhang et al. (2019), Chebana and Ouarda (2021)

Usually, given their several advantages, copula functions are employed in order to model the dependence structure between hydrological variables. On the other hand, the changing environments could alter not only the statistical characteristics of some single random variables (margins) but also the dependence structure between different individual random variables (through copula). Therefore, in addition to studying the nonstationarity in each margin separately, it is also essential to examine the evolution of the dependence structure between those variables via copula. This is possible by studying the nonstationarity in a multivariate context.

Based on the number of variables involved and the stationarity assumption aspect, HFA frameworks could be categorized as follows: (1) univariate stationary, (2) univariate nonstationary, (3) multivariate stationary, and (4) multivariate nonstationary. Table 7.1 provides some references for each framework by focusing on the multivariate nonstationary framework (the topic of this chapter).

7.2 Multivariate nonstationary hydrological frequency analysis literature

First, the nonstationarity, among other assumptions, should be tested in an HFA setting based on univariate and multivariate tests. In general, testing the basic assumptions either in multivariate or univariate frameworks is important. Its results are useful for model selection, especially in the case where the assumptions are not fulfilled. Numerous multivariate tests have been developed and presented in the literature in order to test nonstationarity in a multivariate setting. Mainly, these tests are extensions of the well-known Mann–Kendall and Spearman univariate tests (see also Chapter 4). It is appropriate to consider those tests in multivariate nonstationary modeling.

Numerous studies in statistical hydrology have been conducted on multivariate nonstationarity modeling (see Table 7.1). In this framework, different scenarios can be considered based on the affected variables (e.g., flood volume, peak, or both), the concerned parameters of the model (e.g., location, scale, or dependence), and the changing shapes (e.g., linear or quadratic). In a multivariate HFA, nonstationarity could be present in the statistical characteristics of the margins and/or their dependence structure.

A number of studies have been conducted on the multivariate nonstationary framework, of which some of them are briefly presented here along with more details in the following sections. In a conditional multivariate model based on copulas, Corbella and Stretch (2013) considered nonstationarity only in the margins, whereas the dependence structure was assumed to be constant over time. The idea of employing copulas with time-varying parameters was first introduced in Chebana et al. (2013) to deal with nonconstant dependence structure (Sarhadi et al., 2016). Some studies focused on dependence structures in copula-based modeling under nonstationarity in multivariate HFA using time-varying copulas. In some early studies, the focus was on detecting nonstationarity in the dependence structures as well as in marginal distributions (Ben Aissia et al., 2014; Chebana et al., 2013; Yilmaz et al., 2014). In addition to testing, Ben Aissia et al. (2014) analyzed trends of the dependence between flood characteristics based on moving window series of three dependence measures (Pearson, Kendall, and Spearman coefficients, discussed in Chapter 3). To study the time-dependent behavior of bivariate hydrological design parameters, Bender et al. (2014) presented a bivariate nonstationary approach. To model the peak and volume of flood on the Rhine River, they made use of time-varying parameters both in copulas and in an extreme value distribution (for margins). Bender et al. (2014) showed that compared to the trend in dependence (via the copula parameter), the trends in the margins (through the distribution parameters) are significantly more effective on the corresponding design values. However, this conclusion is not general and could be different for other situations (with other data sets and/or other copula functions) where the trend in the dependence could be larger on the design values. In studying low flows at two locations in China, Jiang et al. (2015) considered a nonstationary copula model where marginal and joint distributions integrated time and reservoir indexes. In contrast to Jiang et al. (2015), Ahn and Palmer (2016) explored the relationship between the bivariate characteristics of low flow focusing on forecasting future bivariate low flow frequency in the Connecticut River Basin, USA.

For an adaptive design framework under nonstationary conditions, Bayesian dynamic conditional copula was developed by Sarhadi et al. (2016). One of their objectives is to model the time-varying dependence structure between mixed continuous and discrete hydrometeorological variables. Their results showed that, under the impact of climate change, the nature and the risk of multivariate extreme-climate processes change over time. They concluded that the long-term decision-making strategies should be updated accordingly. In a similar framework, drought severity and duration over California have been studied by Kwon and Lall (2016). Likewise in a Bayesian framework, Bracken et al. (2018) assumed a Gaussian elliptical copula along with generalized extreme value (GEV) marginal distributions.

As in the univariate framework, indices of large-scale climate drivers such as the El Nino Southern Oscillation (ENSO) and the Pacific Decadal Oscillation (PDO) are also used as covariates (instead of time) to model temporal multivariate nonstationarity. Hence, Zhang et al. (2019), in a multivariate nonstationary Bayesian approach, considered time and climate indices as covariates in the margins while keeping the copula constant. They showed some of the advantages of considering multivariate HFA in terms of uncertainty reduction when estimating return levels as well as in terms of better capturing the dependence structure. In four study regions in eastern coastal China, Li, Wang, Fu et al. (2019) and Li, Wang, Singh et al. (2019) presented a nonstationary HFA of annual extreme rainfall using the variables volume and intensity. When considering the simultaneous occurrence of floods in two (or more) rivers, Feng et al. (2020) performed nonstationary HFA using time-varying copulas for flood coincidence risk analysis. Recently, Chebana and Ouarda (2021), after performing univariate and multivariate trend tests, proposed a dynamic copula model along with univariate nonstationary models (GEV, Log-Normal 2, and Log-Normal 3). In terms of application, they applied it to floods.

7.3 Multivariate nonstationary models

7.3.1 Modeling description

In this section, we focus on presenting the main available models in the multivariate nonstationary HFA framework. Basically, these models vary with parameters in the copula and/or the margins either frequentist (Bracken et al., 2018; Kwon and Lall, 2016; Sarhadi et al., 2016; Zhang et al., 2019) or Bayesian (Bender et al., 2014; Chebana and Ouarda, 2021; Feng et al., 2020, among others). To the best of our knowledge, no comparison is available between these multivariate nonstationarity models yet. However, comparisons between the latter and their stationary counterparts are available, which, in general, show the superiority of the nonstationary ones (e.g., Zhang et al., 2019).

A key ingredient in the multivariate nonstationary modeling is the so-called dynamic copula where the parameters of the copula are not constant (e.g., van den Goorbergh et al., 2005). Before being introduced in the hydro-climatology area, dynamic copula is studied in the fields of finance and tourism to deal with nonstationarity in the dependence economic framework. In multivariate HFA, parametric models, including copulas, are recommended. One of the advantages of parametric models in HFA is their ability to estimate large quantiles in particular for the hydraulic design of major structures.

Let X and Y be two random variables, such as flood peak and volume, with samples denoted by $(x_1, y_1), \ldots, (x_n, y_n)$ of size n. In addition, let F_X and F_Y be the marginal distribution functions, respectively, of X and Y, and F be the joint bivariate distribution function of (X, Y). In order to model (X, Y) in a multivariate nonstationary framework, the parameters of the joint distribution F are considered to be evolving with respect to a given covariate v, where $\theta_C(v), \theta_X(v)$ and $\theta_Y(v)$ represent, respectively, the parameter of the copula $C_{\theta_C(v)}(.,.)$ and the margins $F_{X,\theta_X(v)}(.)\text{and}F_{Y,\theta_Y(v)}(.)$. The covariate $v = (v_1, v_2, \ldots, v_n)$ is assumed to belong to a subset Λ from $\mathbb{R}$, which could be, for instance, the time or a large-scale climate index (e.g., ENSO and PDO). According to Sklar's theorem (Chapter 5), we have

$$F_{\theta(\nu)}(x,y) = C_{\theta_C(\nu)}\left(F_{X,\theta_X(\nu)}(x),\ F_{Y,\theta_Y(\nu)}(y)\right)\ x,y \in \mathbb{R} \text{ and } \nu \in \Lambda \tag{7.1}$$

with $\theta(v) = (\theta_X(v), \theta_Y(v), \theta_C(v))$ being the whole parameter vector. Although we present the case where all parameters are not constant, it is possible that only a part of the parameters is evolving with the covariate.

Commonly, in HFA applications, the parameters of the marginal distribution as univariate distributions are composed of a vector of the location, scale, and shape (such as the case of GEV). However, the considered copulas consist of a single real-valued parameter, which can be expressed in terms of the well-known dependence coefficient measures, such as Kendall's tau τ_K and Spearman's rho ρ_S (see Chapter 5 for expressions and more details). Generally, trends in location measures, such as the mean or median, are the most studied since they are visible and they can be easily detectable. However, even though they have been considered in the literature, usually scale trends are not directly visible and detectable compared to those for location. Finally, when it comes to trends in shape parameter, this parameter is considered constant in the literature. One of the reasons is that a significant number of distrubutions in HFA do not have a shape parameter. Also, the presence of a trend in shape parameter may lead to a complete change in the class of the distributions, such as in the GEV distribution where the distribution could change from Weibull to Fréchet according to the sign of the shape parameter. Regarding the trend in the dependence structure, it is not straightforward to visualize directly from the observations. Indeed, unlike for the margin parameters, the dependence is a notion that entails at least two series. A simple way to exhibit how evolving the dependence structure is through a rolling window series of a given association measure to be plotted with respect to time or covariate (see e.g., Chebana & Ouarda, 2021).

When it comes to evolving parameters with respect to the covariate v, there are various options to deal with this situation. For instance, in the variable X, the trend is linear in the location parameter, whereas in the variable Y, it is the scale parameter that has a quadratic trend, and regarding the trend in the dependence, the parameter is linear. Hence, the corresponding parameters of the joint distribution $\theta_X = (\theta_{X,1}, \ldots, \theta_{X,r})$, $\theta_Y = (\theta_{Y,1}, \ldots, \theta_{Y,s})$ and $\theta_C = (\theta_{C,1}, \ldots, \theta_{C,\ell})$ can be expressed as $\theta_{X,j}(\nu) = \alpha_{0,j} + \alpha_{1,j}\nu + \alpha_{2,j}\nu^2$, $\theta_{Y,j}(\nu) = \beta_{0,j} + \beta_{1,j}\nu + \beta_{2,j}\nu^2$ and $\theta_{C,j}(\nu) = \gamma_{0,j} + \gamma_{1,j}\nu + \gamma_{2,j}\nu^2$ for $j = 1, \ldots, r$ or s or ℓ. As discussed above, for HFA applications, r and s could be 2 or 3 wheras ℓ is 1. The constants α, β and γ are called the hyperparameters where some of them could be null, leading to a stationary, linear, or quadratic trend. For example, Li, Wang, Fu et al. (2019) and Li, Wang, Singh et al. (2019) employed time-dependent Archimedean copulas and GEV margin distributions.

7.3.2 Covariate-varying copulas

In the multivariate HFA, several copulas are considered, in particular but not limited to the Archimedean or Extreme Value families (see Chapter 5), such as the following:

1. The Gumbel copula (Archimedean) is given by

$$C_\gamma(u,v) = \exp\left(- \left[(-\log\ u)^\gamma + (-\log\ v)^\gamma\right]^{\frac{1}{\gamma}}\right) \text{ for } u, v \in [0,1] \text{ and } \gamma \geq 1 \tag{7.2}$$

2. The Galambos copula (Extreme Value) is expressed as

$$C_\gamma(u,v) = uv\ \exp\left(\left[(-\log\ u)^{-\gamma} + (-\log\ v)^{-\gamma}\right]^{\frac{-1}{\gamma}}\right) \text{ for } u, v \in [0,1] \text{ and } \gamma \geq 0 \tag{7.3}$$

3. The Hüsler–Reiss copula (Extreme Value) is given by

$$C_\gamma(u,v) = \exp\left[(\log\ u)\Phi\left(\frac{1}{\gamma} + \frac{\gamma}{2}\log\left(\frac{\log\ u}{\log\ v}\right)\right) + (\log\ v)\Phi\left(\frac{1}{\gamma} + \frac{\gamma}{2}\log\left(\frac{\log\ v}{\log\ u}\right)\right)\right] \tag{7.4}$$

for $u, v \in [0, 1]$ and $\gamma \geq 0$, where the function Φ (.) represents the cumulative distribution function of the standard normal distribution.

In general, the parameters of a given model should consider specific constraints such as belonging to an interval, being positive, or different from zero. Because in the (multivariate) nonstationary framework the parameters are evolving and not constant, it is necessary to ensure that those parameters although evolving should consider the constraints of the associated models. To this end, in some situations, appropriate transformations could be required to be considered in such a model. One of the typical examples is regarding the scale parameter, which should be always positive. Hence, an appropriate transformation to ensure considering this constraint is an exponential transformation. In the case of Gumbel copula, the parameter γ should be greater than 1, and we use the transformation $\varphi_\gamma(v) = \log(\gamma(v) - 1)$. However, for Galambos and Hüsler–Reiss copulas, in order to keep the parameter positive, we consider the transformation $\varphi_\gamma(v) = \log\ \gamma(v)$.

Note that it is not recommended to consider complex or unrealistic trend shapes because of the parsimony principle involved in modeling. Usually, for instance, regarding Gumbel copulas, there are three cases of trend related to the variation of its parameter γ: stationary, linear nonstationary, and quadratic nonstationary with respect to the covariate v corresponding, respectively, to Gumbel$_0$ (no trend $\gamma(v) = 1 + \exp(\gamma_1)$), Gumbel$_1$ (linear trend $\gamma(v) = 1 + \exp(\gamma_1 + \gamma_2 v)$), and Gumbel$_2$ (quadratic trend $\gamma(v) = 1 + \exp(\gamma_1 + \gamma_2\ v + \gamma_3 v^2)$), where γ_1, γ_2 and γ_3 are the hyperparameters. Similar notations can be defined for other copulas such as the above Galambos and Hüsler–Reiss copulas.

7.3.3 Covariate-varying margins

Since in HFA the focus is on extreme events, the joint distribution (including marginal distributions and copulas) should be selected accordingly (see Chapter 5). In univariate nonstationary HFA, among the most employed distributions, we have the GEV and log-Normal distributions with two or three parameters (denoted LN2 and LN3, respectively). For GEV, recall that its cumulative distribution function is given by

$$F_{\mathrm{GEV}}(x; \mu, \sigma, k) = \exp\left[-\left(1 + k\left(\frac{x-\mu}{\sigma}\right)\right)^{-\frac{1}{k}}\right] \quad \text{if } k \neq 0 \text{ and } \exp\left[-\exp\left(\frac{-(x-\mu)}{\sigma}\right)\right] \quad \text{if } k = 0 \tag{7.5}$$

for $x \geq \left(\mu - \frac{\sigma}{k}\right)$ if $k > 0$ (Fréchet), $x \in \mathbb{R}$ if $k = 0$ (Gumbel) and $x \leq \left(\mu - \frac{\sigma}{k}\right)$ if $k < 0$ (Weibull) where $\mu \in \mathbb{R}$, $\sigma > 0$ and $k \in \mathbb{R}$ are the location, scale, and shape parameters, respectively. However, for LN2 and LN3 distributions, it is preferable to present their density functions, respectively, as

$$f_{LN2}(x;\mu,\sigma)=\frac{1}{x\sigma\sqrt{2\pi}}e^{-\frac{(\log x-\mu)^2}{2\sigma^2}}\ \text{for}\, x>0 \tag{7.6}$$

$$f_{LN3}(x;\mu,\sigma,m)=f_{LN2}(x-m;\mu,\sigma)\ \text{for}\, x>m \tag{7.7}$$

where $\mu\in\mathbb{R}$ and $\sigma>0$ are, respectively, the mean and standard deviation of $\log x$, whereas $m\in\mathbb{R}$ is a threshold parameter (lower bound).

The considered nonstationary GEV version $GEV(\mu(v),\sigma(v),k)$ incorporates nonstationarity in its location and scale parameters by linking them to the covariate v. As mentioned above, for simplicity and for practical reasons, the shape parameter k is considered constant. According to the shape of the trend, we consider the notations GEV_{00}: no trend; GEV_{10}: linear trend in the location parameter $\mu(v)=\mu_0+\mu_1 v$; GEV_{20}: quadratic trend in the location parameter $\mu(v)=\mu_0+\mu_1 v+\mu_2 v^2$; and GEV_{11}: linear trend in both the location and scale parameters $\mu(v)=\mu_0+\mu_1 v$, $\sigma(v)=\exp(\sigma_0+\sigma_1 v)$. As mentioned earlier, to keep the scale parameter $\sigma(v)$ positive, we use the transformation $\varphi_\sigma(v)=\log(\sigma(v))$. Using similar notations, we obtain the models $LN2_{00}$, $LN2_{10}$, and $LN2_{20}$ in analogy with the location parameter μ, respectively, as in GEV_{00}, GEV_{10}, and GEV_{20}. For the LN3 case, the models $LN3_{00}$, $LN3_{10}$, $LN3_{20}$, and $LN3_{11}$ can be considered as the possible trend in threshold parameter m, that is, $LN3(\mu(v),\sigma,m(v))$.

7.3.4 Nonstationary model selection

Usually, in the HFA nonstationarity context, selection criteria such as the Akaike information criterion (AIC) or the Bayesian information criterion (BIC) are employed to select the appropriate distributions. Indeed, in this context, the number of possible models could be very large, which increases with the number of the involved variables, the affected parameters, and the form of the effects. In the univariate case of the GEV model as an example, we have two parameters possibly affected (location and scale) with three options of the trend (constant, linear, or quadratic) leading to six models to be compared. In the multivariate nonstationary case, the number of models to be compared could become very large, where in a usual bivariate nonstationary example, we have 15 models (e.g., six models for each of two margins and three models for the copula). Therefore, it is appropriate to use general criteria like the AIC. The latter is already used in the univariate nonstationary modeling in hydrology as well as in the multivariate nonstationary framework with copula for financial applications. Note that in the univariate nonstationarity framework, the deviance statistic is utilized as a model-selection criterion among a group of *nested* models, that is, compare a model M1 with its sub-model M0 (El Adlouni et al., 2007). However, considering only nested models in the multivariate nonstationary context is not reasonable and very restrictive.

7.3.5 Bayesian multivariate nonstationary model

In order to model a time-varying dependence structure, Sarhadi et al. (2016) and Kwon and Lall (2016), among other studies metioned earlier, developed a Bayesian dynamic conditional copula. In a joint Bayesian inference, to fit the margins and copula, an adaptive Gibbs Markov chain Monte Carlo (MCMC) sampler is employed. For the model parameters, posterior mean estimates and credible intervals are obtained. In addition, to select the appropriate nonstationary model, the deviance information criterion (DIC) is considered.

All the model parameters are estimated with a full likelihood-based Bayesian inference. More specifically, in Sarhadi et al. (2016), marginal distribution parameters as well as copula parameters are expressed as functions of time through the generalized regression linear model. For instance, in their case of droughts, the parameters are the time-varying distribution parameters $\mu_t, \lambda_t, \theta_{ct}$ and γ_t associated, respectively, with drought severity, duration, and interarrival time and the copula parameters. The generalized linear model parameters, say, $\beta_S, \beta_D, \beta_A$ and β_C, respectively, associating the parameters $\mu_t, \lambda_t, \theta_{ct}$ and γ_t with time t as a covariate, are estimated with Bayesian inferences.

In the Bayesian inference framework, prior distributions are identified for all the unknown parameters of the generalized linear model. Next, the observed data are combined with the information obtained from the prior distributions leading to posterior distributions. In the case of Sarhadi et al. (2016), a two-step Bayesian approach is proposed. In the first step, the inference is for the marginal parameters β_S, β_D *and* β_A, whereas it is for the copula parameter β_C in the second step based on the results of the first step.

In the Bayesian framework, the prior distributions play a role in providing any prior information regarding the parameters, $\beta_S, \beta_D, \beta_A$ and β_C. In their example, proper but weakly informative priors are assumed for each parameter, such as Normal or Gamma distributions. The reader is referred to Sarhadi et al. (2016) or Kwon and Lall (2016) for more details and explicit formulations.

In general, frequencist methods are usually preferred for estimating parameters, such as in HFA. However, the Bayesian framework could have some attractive advantages for time-varying copula estimation, as indicated in Sarhadi et al. (2016).

7.3.6 Modeling procedure steps

The main steps of this procedure are discussed here in more detail. A first preliminary step consists of a descriptive statistical analysis for each of the involved variables as well as evaluation of the different association measures. In a second step before modeling, it is required to conduct trend tests (univariate and multivariate). Regarding trend detection in the dependence structure between the series, rolling window series from the association measures can be used. The next step deals with selecting the possibly nonstationary marginal distributions of X and Y with their estimated parameters. In addition, the most appropriate copula (stationary or not) should be selected along with estimating its parameters. Based on the results of the previous steps, stationary or nonstationary multivariate quantiles and return periods can be obtained.

Step 1: descriptive study

The aim of the first step is to highlight the marginal characteristics of the series $x_1, \ldots, x_n$ and $y_1, \ldots, y_n$, such as presenting graphics of the series with respect to the covariate v. These graphs are useful to visually identify if the series ($x_1, \ldots, x_n$ and $y_1, \ldots, y_n$) have possible marginal trends.

Step 2: testing trends

Testing the basic assumptions in HFA is presented in Chapter 4, including multivariate trend testing. Several tests can be applied on both univariate and multivariate series. In addition, the different association measures should be evaluated between the original series $x_1, \ldots, x_n$ and $y_1, \ldots, y_n$ over a rolling window for q windows of size s $(s < n)$ to obtain a series for each association measure. The obtained series can be plotted and may allow us to detect graphically any possible presence of a trend in the dependence. Furthermore, a stationarity test can be applied on the rolling window series for each of the association measures. However, since these series are auto-correlated by construction, the usual Mann–Kendall trend test cannot be applied directly. Hence, appropriate versions of this test can be considered such as the prewhitening, the trend-free prewhitening, and block bootstrap. Given some of its practical benefits, the block bootstrap approach is preferred. However, Li, Wang, Fu et al. (2019) and Li, Wang, Singh et al. (2019) considered the prewhitening version of the test.

Step 3: joint distribution selection

Based on Sklar's theorem (see Chapter 5), the joint distribution is composed of a margin for each involved variable along with a copula. Hence, this step is divided into the following two parts.

3a: selection of nonstationary margins

In order to identify the most appropriate nonstationary marginal distribution for each variable, we consider the widely used distributions in univariate HFA. In the present context, we have GEV, LN2, and LN3 distributions. To estimate the corresponding hyperparameters, in general, the maximum-likelihood (ML) approach can be employed as well as specific methods for a given distribution such as the generalized ML for the GEV. Then, the AIC criterion is evaluated for all considered models and the selected model is the one with the lowest AIC.

3b: selection of nonstationary copula

To select the appropriate copula, the same principle of the classical case (presented in Chapter 5) is valid. First, it is of interest to restrict attention to potential classes of copulas that are commonly appropriate for hydrological applications and in particular in HFA, such as those with upper-tail dependence (e.g., Gumbel, Galambos, and Hüsler–Reiss). For each one of the above copulas, the corresponding hyperparameters $(\gamma_1, \gamma_2, \gamma_3)$ are estimated using the maximum pseudo-likelihood (MPL) approach and the appropriate transformation ϕ_γ when needed. Then, the AIC criterion is evaluated for each of the considered copula models, including their nonstationary versions.

Note that the MPL is preferred as an estimation method for copula parameters (see Chapter 5). The adopted AIC version is based on the MPL instead of the usual ML. Hence, the corresponding AIC expression is given by

$$\text{AIC} = -2\log L_n(\hat{\gamma}) + 2h \tag{7.8}$$

where $L_n(\hat{\gamma})$ stands for the value of the pseudo-likelihood $L_n(\gamma)$ at the MPL estimator $\hat{\gamma}$ and h is the total number of all the involved parameters. The pseudo-likelihood $L_n(.)$ is given by

$$L_n(\gamma) = \prod_{j=1}^{n} c_\gamma\left(u_{1j}, u_{2j}; \gamma\left(v_j\right)\right) \tag{7.9}$$

where c_γ stands for the density function corresponding to the copula C_γ, and u_{ij} are the pseudo-observations given by

$$u_{ij} = \frac{r_{ij}}{(n+1)} \quad \text{for } i = 1, 2 \text{ and } j = 1, \ldots, n \tag{7.10}$$

where r_{ij} is the ascending ranks associated with the samples of X and Y, respectively (see Chapter 5).

Step 4: nonstationary multivariate quantile and return period

Once the appropriate joint distribution (copula and margins) is selected as well as the corresponding parameters estimated, the last step consists of getting the corresponding nonstationary multivariate quantile given a selected value $0<p<1$ (see Chapter 6). Explicitly, the bivariate nonstationary quantile is given by

$$Q_{X,Y}(p,v)=\left\{(x,y)\in R^2 \text{where} x=F^{-1}_{X,\theta_X(v)}(u), y=F^{-1}_{Y,\theta_Y(v)}(v); u,v\in[0,1]: C_{\gamma(v)}(u,v)=p\right\} \tag{7.11}$$

In this case, $Q_{X,Y}(p,.)$ represents a collection of quantile curves for the same given risk p, where each curve is associated with a fixed value v of the covariate.

The overall shape and behavior of this quantile curve collection may change according to the component of the joint selected distribution (one or more of the margins and the copula are affected by a trend and its form, e.g., linear or quadratic). Fig. 7.1 illustrates some of those situations. This illustration shows that the variation of the curves is most affected, especially in the naïve part, when the trend is on the margins (all sub-figures *except* Fig. 7.1A and D). However, the proper part of the curves is more affected when a trend is in the copula parameter (Fig. 7.1D, F–H). Even though these are only illustrations and cannot be generalized, they indicate that disregarding trends in some or all components of joint distribution parameters can conduct to misleading conclusions in terms of safety and cost. Indeed, the obtained quantiles may not correspond to the actual risk.

As presented in Chapter 6, in the multivariate HFA framework, a variety of options and definitions regarding the multivariate return period (RP) have been suggested in the literature. These options and definitions can be extended to the multivariate nonstationary case. For instance, fully time-varying RP is proposed by Sarhadi et al. (2016). In their study about droughts, they defined the time-varying RP at time t as the expected time between droughts with duration larger than d_0 and severity larger than s_0 at time t:

$$\mathrm{RP}_t(d_0,s_0)=\frac{E(X_t)}{1-F_{D_t}(d_0)-F_{S_t}(s_0)+P(D_t\le d_0,S_t\le s_0)}$$

where $E(X_t)$ is the mean interarrival time between drought events. This formulation can be adapted to other hydrological events such as floods (e.g., Chebana & Ouarda, 2021) or other RP types such as Kendall's return period (e.g., Li, Wang, Fu et al., 2019; Li, Wang, Singh et al., 2019). The above formulations and illustrations indicate that the copula parameter could change with time or covariate, leading to changes in the quantile curves and the RP as well. Hence, it is important to not ignore the

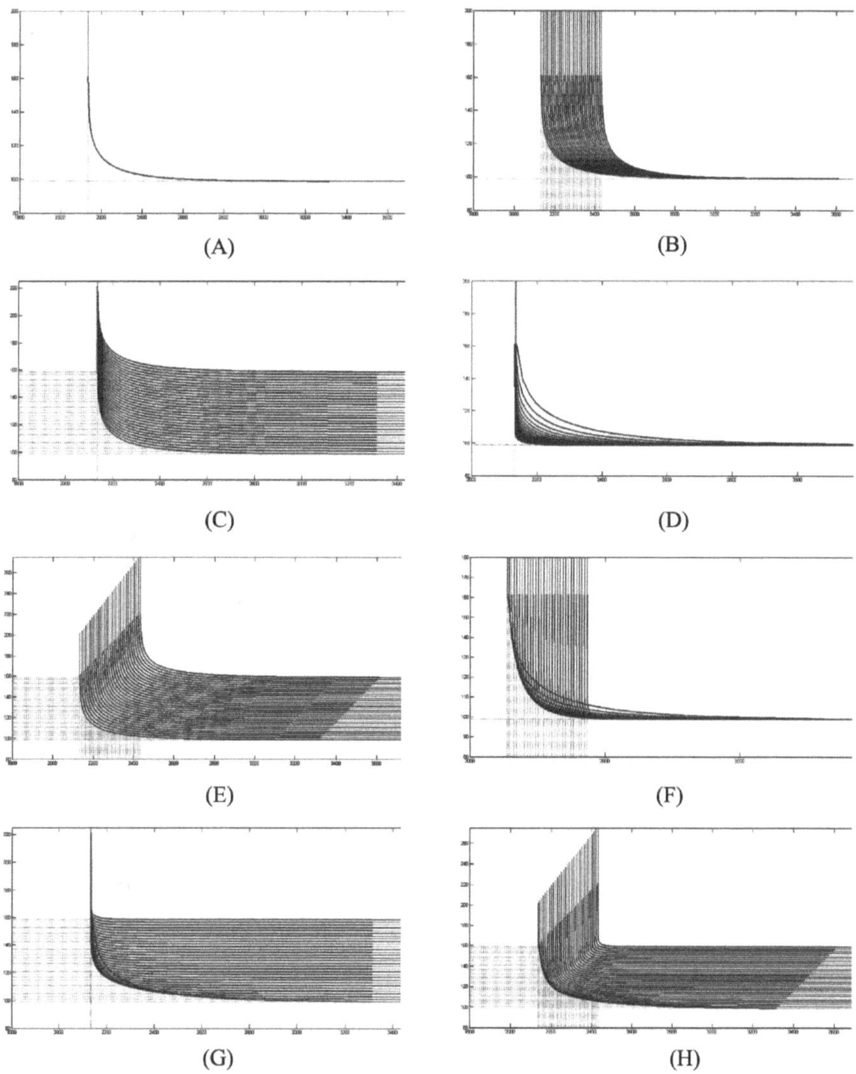

FIGURE 7.1 Illustration of different trend options of the bivariate quantile collection curves. They correspond to the event nonexceedance for both variables (Chebana & Ouarda, 2021). (A) All stationary (no trend); (B) only linear trend in X; (C) only linear trend in Y; (D) only linear trend in copula parameter; (E) linear trend in both X and Y; (F) linear trend in both X and copula parameter; (G) linear trend in both Y and copula parameter X; (H) linear trend in X, Y and copula parameter.

nonstationarity aspect in the dependence structure. In conclusion, to be more realistic and to obtain more accurate risk estimations, quantiles and RP should integrate possible nonstationarity in multivariate HFA from the early stages of the analysis.

7.3.7 Data series length and moving window series

The notion of nonstationarity, especially trend, is closely related to long-term and long data series either in univariate HFA (e.g., Salas & Obeysekera, 2014) or in multivariate HFA (e.g., Li, Wang, Fu et al., 2019; Li, Wang, Singh et al., 2019). Hence, in studying nonstationarity, long records of data series are commonly required both for statistical and for hydro-climatological reasons. On the one hand, multivariate HFA requires more data than its univariate counterpart, and on the other hand nonstationary analysis requires more data compared to the stationary case. Hence, record length becomes more important when combining both multivariate and nonstationarity frameworks. Behind the hydro-climatological reasons, one of the statistical reasons being the rapid increase in the number of parameters and hyperparameters according to the number of involved variables (dimension) and the complexity of the trend (number of affected components and shape of trends).

Commonly in HFA, data sets are relatively short. Therefore, in order to improve copula modeling as well as to perform nonstationary analyses, a feasible approach to enrich the databases is required. To this end, the concept of compatibility has been proposed by Grimaldi et al. (2016). Essentially, to increase the sample size, two or more multivariate "compatible" data sets can be merged. This approach appears to be effective for evaluating the degree of compatibility between catchments in terms of their dependence structure and then for increasing the data set, which potentially improves multivariate HFA estimations.

Given a sample size n, and a window of width s, the size q of the rolling window series is $q = n - s + 1$. Choosing a value of s should be seen as a balance between small and large values similar to other situations in statistics such as the window in nonparametric density estimation. Small values of s increase the number of rolling window series q. This ensures reliability of the analysis and the corresponding results. However, large values of s would decrease q but allow us to obtain efficient values of the dependence coefficients in each window. As indicated above, given the relatively short sample sizes in HFA, this concern is challenging, but the situation is expected to improve since in the future sample size would increase (Chebana & Ouarda, 2021). As an example, in a nonstationarity analysis, Li, Wang, Fu et al. (2019) and Li, Wang, Singh et al. (2019) applied a moving time window of 30 years to the index series.

7.4 Illustrative example

In this example, we consider the dataset used in Chapter 4 where one or more trends have been detected. It is based on the data of the Moisie station located in the province of Quebec, Canada.

Recall that from Chapter 4, the peak Q and volume V series are found to have significant trends at $\alpha = 0.05$. In addition, for the pair (Q, V) series, all the considered tests suggest rejecting the null hypothesis where a trend is significant. However, note that the p-values of the multivariate Mann–Kendall CST and CIT tests are relatively close to the nominal values $\alpha = 0.05 \text{or} 0.01$ (respectively, 0.04 and 0.003 for MK tests and 0.03 and 0.002 for Spearman tests). This could be seen as not clear rejection of H_0.

Note that in this example, the covariate v is time in units of years. Given the sample size $n = 35$, by choosing a rolling window size of $s = 16$ years, the rolling window series is of size $q = 20$. As discussed above, this number is good enough in an HFA context to obtain relevant series of ρ_P, τ_K and ρ_S to perform the analysis. Fig. 7.2 presents the obtained rolling windows series of τ_K.

We observe an increase in τ_K in the few first years and a decrease in the late years, but most of the series points, in the middle period, are around the empirical value (blue line in Fig. 7.2) of $\tau_K = 0.52$ with the whole data. Overall, despite the decrease since early 80s, it is still not a strong trend. However, this situation should be analyzed again when more data will be available since there is a potential of a decreasing trend in the dependence between Q and V.

Regarding the margins, both for Q and V, the GEV, LN2, and LN3 distributions are considered covering for each distribution stationarity as well as different forms of nonstationarity. The corresponding AIC and the estimated parameters and hyperparameters are summarized in

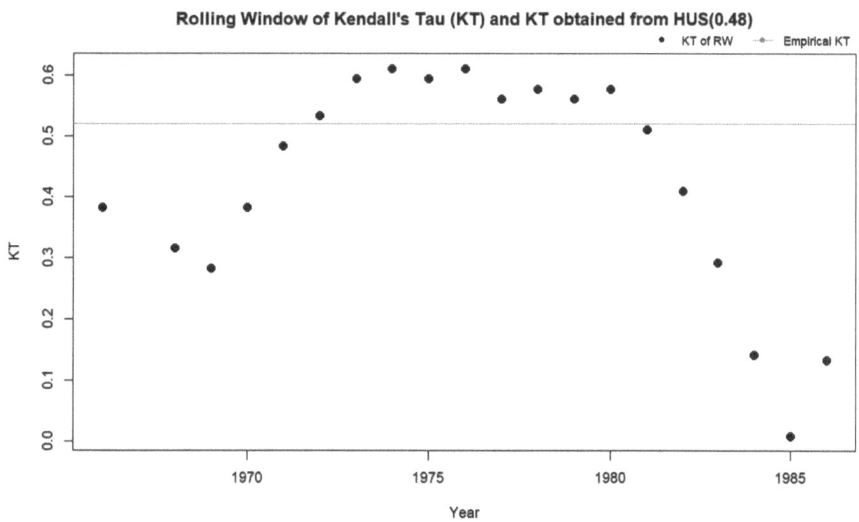

FIGURE 7.2 The rolling window series for τ_K between V and Q for the Moisie data set.

TABLE 7.2 Akaike information criterion (AIC) for the margins of the considered models, as well as the estimated hyperparameters.

Volume	**AIC**	μ_0	μ_1	μ_2	σ_0	σ_1	k_0	k_1
GEV_{00}	610.8777	4661.9065			1293.2568		−0.1935	
GEV_{10}	600.9219	6005.9500	−69.0209		1020.9917		−0.0683	
GEV_{20}	603.8787	6288.6874	−102.3796	0.8729	1189.0679		−0.1335	
GEV_{11}	602.8355	6006.9129	−68.6886		948.2182	4.1878	−0.0824	
$LN2_{00}$	609.2992	8.5213			0.2744			
$LN2_{10}$	**600.4866**	**8.7792**	**−0.0134**		**0.2351**			
$LN2_{20}$	602.2581	8.8279	−0.0203	0.0002	0.2344			
$LN3_{00}$	611.1979	8.8340			0.2004		−1792.5315	
$LN3_{10}$	602.5126	8.8308	−0.0133		0.2233		−260.6803	
$LN3_{20}$	604.2605	8.8329	−0.0211	0.0002	0.2345		4.0032	
$LN3_{11}$	604.4890	8.8579	−0.0122		0.2122		−527.3963	0.0260
$LN3_{21}$	606.2577	8.8326	−0.0202	0.0002	0.2326		−35.6092	−0.0002
Peak								
GEV_{00}	548.2853	1962.1948			511.1820		−0.1205	
GEV_{10}	537.3707	2872.3401	-41.3323		521.6979		−0.4765	
GEV_{20}	549.4960	1958.2850	14.3955	−0.6848	477.4598		−0.1375	
GEV_{11}	**533.7664**	**2354.5388**	**−15.8738**		**778.8625**	**−14.7532**	**−0.4389**	
$LN2_{00}$	546.6041	7.6632			0.2643			
$LN2_{10}$	537.0782	7.9184	−0.0133		0.2242			
$LN2_{20}$	537.3418	8.0450	−0.0313	0.0005	0.2187			
$LN3_{00}$	548.3755	7.9388			0.2000		−657.7194	
$LN3_{10}$	538.8685	7.9422	−0.0118		0.2111		−117.7786	
$LN3_{20}$	540.1217	7.9424	−0.0318	0.0005	0.2444		197.7103	
$LN3_{11}$	539.5182	8.6714	−0.0059		0.0896		−3050.5795	0.6948
$LN3_{21}$	541.9333	7.9436	−0.0247	0.0003	0.2294		80.6794	0.0012

The bold values indicate the smallest Akaike information criterion values and the selected model.

Table 7.2. Therefore, for the flood volume V, the model $LN2_{10}$ is the one with the lowest AIC value and hence can be considered as the most appropriate model for V. Similarly, regarding the peak flood Q, it is GEV_{11}. This model selection is in agreement with the Mann–Kendall trend test results in Chapter 4.

In analogy to Table 7.2, AIC values as well as the estimated parameters and hyperparameters of all considered copulas are presented in Table 7.3. Accordingly, the copula corresponding to the lowest AIC value is the stationary Hüsler–Reiss copula HUS_0. The next choices are still stationary copulas (GUM_0 and GAL_0). We observe that the values of the hyperparameters γ_1 and γ_2 are very small, which supports the choice of a stationary copula.

In conclusion, the selected bivariate joint distribution is composed of nonstationary margins as well as a stationary copula. This is consistent with the trend testing results showing a significant trend for both peak and volume, whereas a slight trend is demonstrated with the multivariate tests. Table 7.4 summarizes the selected bivariate distribution (margins and copula) as well as the estimated parameters and hyperparameters.

Based on the selected bivariate distribution, summarized in Table 7.4, the corresponding bivariate quantile curves are presented in Fig. 7.3 for p

TABLE 7.3 Akaike information criterion (AIC) values and the estimated hyperparameters for the considered copula models.

Copula model	**AIC**	γ_0	γ_1	γ_2
GUM_0	−15.8385	−0.1286		
GUM_1	−13.8388	−0.1325	0.0002	
GUM_2	−13.0842	−0.8865	0.1101	−0.0028
GAL_0	−15.8422	0.1379		
GAL_1	−13.8567	0.0976	0.0023	
GAL_2	−13.0423	−0.4019	0.0779	−0.0019
HUS_0	**−15.9120**	**0.4765**		
HUS_1	−14.0346	0.3976	0.0049	
HUS_2	−12.8214	0.0947	0.0499	−0.0012
JOE_0	−11.0878	0.1116		
JOE_1	-9.3064	−0.0926	0.0128	
JOE_2	−8.0019	−0.8400	0.1073	−0.0022

GUM, Gumbel copula; *GAL*, Galambos copula; *HUS*, Hüsler–Reiss copula.
The bold values indicate the smallest Akaike information criterion values and the selected model.

TABLE 7.4 Values of the estimated hyperparameters for the selected bivariate distribution.

	Best model	γ_0	μ_0	μ_1	σ_0	σ_1	k_0
Volume	$LN2_{10}$		8.78	−0.01	0.23		
Peak	GEV_{11}		2354.54	−15.87	778.86	−14.75	−0.44
Copula	HUS_0	0.48					

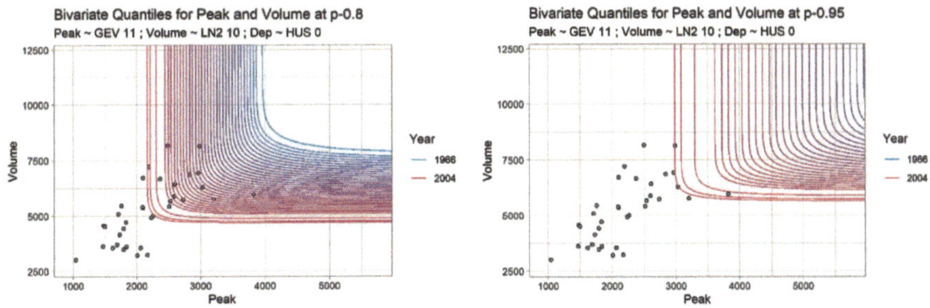

FIGURE 7.3 Bivariate (Q, V) estimated quantile curves corresponding to $p =$ 0.8 and 0.95 (left, right espectively), using the selected models $LN2_{10}$ and GEV_{11}, respectively, for Q and V and Hüsler–Reiss copula.

= 0.80 and 0.95. For both values of p, we observe that the curves are parallel to each other since the selected copula is stationary (as illustrated in Fig. 7.1E). In addition, since both margins are affected by a trend, the curves are well separated at each axe representing Q or V. If one of the margins is stationary, then the quantile curves would be merged in one single curve in the part of that margin (as illustrated in Fig. 7.1B or C). However, as in the univariate case, the quantile curves are "higher" for $p = 0.95$ than for $p = 0.8$. Indeed, the curves are far away from the data and closer to the top-right corner for $p = 0.95$ compared to those of $p = 0.8$.

References

Ahn, K. H., & Palmer, R. N. (2016). Use of a nonstationary copula to predict future bivariate low flow frequency in the Connecticut river basin. *Hydrological Processes, 30*(19), 3518–3532.

Aissia, M. A. B., Chebana, F., Ouarda, T. B., Roy, L., Bruneau, P., & Barbet, M. (2014). Dependence evolution of hydrological characteristics, applied to floods in a climate change context in Quebec. *Journal of Hydrology, 519*, 148–163.

Bender, J., Wahl, T., & Jensen, J. (2014). Multivariate design in the presence of nonstationarity. *Journal of Hydrology, 514*, 123–130.

Bracken, C., Holman, K. D., Rajagopalan, B., & Moradkhani, H. (2018). A Bayesian hierarchical approach to multivariate nonstationary hydrologic frequency analysis. *Water Resources Research, 54*(1), 243–255.

Chebana, F., & Ouarda, T. B. (2021). Multivariate nonstationary hydrological frequency analysis. *Journal of Hydrology, 593*, 125907.

Chebana, F., Ouarda, T. B. M. J., & Duong, T. C. (2013). Testing for multivariate trends in hydrologic frequency analysis. *Journal of Hydrology, 486*, 519–530.

Corbella, S., & Stretch, D. D. (2013). Simulating a multivariate sea storm using Archimedean copulas. *Coastal Engineering, 76*, 68–78.

El Adlouni, S., Ouarda, T. B. M. J., Zhang, X., Roy, R., & Bobée, B. (2007). Generalized maximum likelihood estimators for the nonstationary generalized extreme value model. *Water Resources Research, 43*(3).

Feng, Y., Shi, P., Qu, S., Mou, S., Chen, C., & Dong, F. (2020). Nonstationary flood coincidence risk analysis using time-varying copula functions. *Scientific Reports, 10*(1), 1–12.

François, B., Schlef, K. E., Wi, S., & Brown, C. M. (2019). Design considerations for riverine floods in a changing climate–a review. *Journal of Hydrology, 574*, 557–573.

Gilroy, K. L., & McCuen, R. H. (2012). A nonstationary flood frequency analysis method to adjust for future climate change and urbanization. *Journal of Hydrology, 414*, 40–48.

Grimaldi, S., Petroselli, A., Salvadori, G., & Michele, C. D. (2016). Catchment compatibility via copulas: A non-parametric study of the dependence structures of hydrological responses. *Advances in Water Resources, 90*, 116–133.

Jiang, C., Xiong, L., Xu, C. Y., & Guo, S. (2015). Bivariate frequency analysis of nonstationary low-flow series based on the time-varying copula. *Hydrological Processes, 29*(6), 1521–1534.

Kwon, H. H., & Lall, U. (2016). A copula-based nonstationary frequency analysis for the 2012–2015 drought in California. *Water Resources Research, 52*(7), 5662–5675.

Li, T., Wang, S., Fu, B., & Feng, X. (2019). Frequency analyses of peak discharge and suspended sediment concentration in the United States. *Journal of Soils and Sediments*, 1–12.

Li, H., Wang, D., Singh, V. P., Wang, Y., Wu, J., Wu, J., ... Zhang, J. (2019). Nonstationary frequency analysis of annual extreme rainfall volume and intensity using Archimedean copulas: A case study in eastern China. *Journal of Hydrology, 571*, 114–131.

Rao, A. R., & Hamed, K. H. (2000). Flood frequency analysis. *New directions in civil engineering*. Boca Raton, FL: CRC Press.

Salas, J. D., & Obeysekera, J. (2014). Revisiting the concepts of return period and risk for nonstationary hydrologic extreme events. *Journal of Hydrologic Engineering, 19*(3), 554–568.

Salas, J. D., Obeysekera, J., & Vogel, R. M. (2018). Techniques for assessing water infrastructure for nonstationary extreme events: A review. *Hydrological Sciences Journal, 63*(3), 325–352.

Sarhadi, A., Burn, D. H., & Wiper, M. P. (2016). Time-varying nonstationary multivariate risk analysis using a dynamic Bayesian copula. *Water Resources Research, 52*(3), 2327–2349.

Strupczewski, W. G., Singh, V. P., & Feluch, W. (2001). Nonstationary approach to at-site flood frequency modelling I. Maximum likelihood estimation. *Journal of Hydrology, 248*(1), 123–142.

van den Goorbergh, R. W. J., Genest, C., & Werker, B. J. M. (2005). Bivariate option pricing using dynamic copula models. *Insurance: Mathematics and Economics, 37*(1), 101–114.

Wang, X., Huang, G., & Liu, J. (2014). Projected increases in intensity and frequency of rainfall extremes through a regional climate modeling approach. *Journal of Geophysical Research: Atmospheres, 119*(23), 13271–13286.

Yilmaz, A. G., Hossain, I., & Perera, B. J. C. (2014). Effect of climate change and variability on extreme rainfall intensity–frequency–duration relationships: A case study of Melbourne. *Hydrology and Earth System Sciences, 18*(10), 4065–4076.

Zhang, X., Duan, K., & Dong, Q. (2019). Comparison of nonstationary models in analyzing bivariate flood frequency at the Three Gorges Dam. *Journal of Hydrology*, 124208.

CHAPTER

8

Multivariate regional frequency analysis

8.1 Regional hydrological frequency analysis

As presented in the previous chapters, hydrological frequency analysis (HFA) can be in different frameworks. Indeed, according to the number of involved variables, it can be either univariate or multivariate. However, depending on the number of considered sites and the available data, this analysis can also be either local (at-site) or regional. Consequently, four frameworks are obtained: (1) univariate local, (2) multivariate local, (3) univariate regional, and (4) multivariate regional. An impressive literature is focused on both univariate frameworks (local and regional) where we can refer for instance to Stedinger and Tasker (1986), Burn (1990), Hosking and Wallis (1997), Wazneh et al. (2013), and Haddad and Rahman (2020). As discussed in previous chapters, the multivariate approach allows improvement of the analysis by considering the dependence between variables and by employing more available information. Consequently, more attention has recently been paid to at-site multivariate framework (2) (see references given in previous Chapters). However, dealing simultaneously with both multivariate and regional aspects is very recent. It corresponds to framework (4) where we can find a number of studies such as Chebana and Ouarda (2007), Sadri and Burn (2012), Ben Aissia et al. (2015), and Requena et al. (2016).

In Chapter 2, the multivariate HFA framework was presented with a number of arguments and advantages for its use. As a summary, (1) it allows for a better understanding of the probabilistic features of extreme events in order to the study their joint distribution, (2) multivariate HFA requires more data and appropriate statistical analysis, (3) univariate

DOI: https://doi.org/10.1016/B978-0-323-95908-7.00008-6

(single-variable) HFA can only provide limited assessment of extreme events, (4) univariate HFA can be valuable in some situations such as when only a single variable is significant for design purposes or when the two or more variables are significantly independent. However, if the dependence/correlation is an essential information in risk assessment, then one or more separate analysis for each single random variable is not able to reveal the relationship between those variables. Hence, it is recommended to *jointly* study the variables characterizing the hydrological event in the multivariate HFA framework.

One of the challenges in HFA is the limitation of available data where a majority of gauging stations have record lengths in the range 30–50 years. Therefore estimations of high return levels could be negatively impacted. To overcome this limitation, regional frequency analysis (RFA) is one of the considered solutions. Indeed, the aim of RFA is to provide estimates of hydrological characteristics at *ungauged* or *partially gauged* sites. RFA attempts to improve at-site quantile estimates at gauged sites, particularly when the available series are short. In addition, RFA provides quantile estimates at ungauged sites where no hydrological data are available. RFA is based on combining the information from a group of stations forming a region. By considering a region, the sample size is increased in RFA which allows to reduce the uncertainty in extreme value estimation. The idea behind RFA can be seen as trading space (region) for time (sample size). Fig. 8.1 provides an illustration of RFA framework.

In general, RFA consists of two main steps: (1) *delineation* of hydrological homogeneous regions and (2) *regional estimation*. Briefly, in the first step, regions are formed by gathering sites based on their hydrologic similarity and then in the second step, hydrological information can be transferred to target sites. The hydrologic similarity of a group of sites should lead to a homogeneous region. To evaluate the later, L-moment based tests have been proposed (Hosking & Wallis, 1997). When it comes to regional estimation (step 2), different methods have been proposed, where regressive models and the index-flood methods

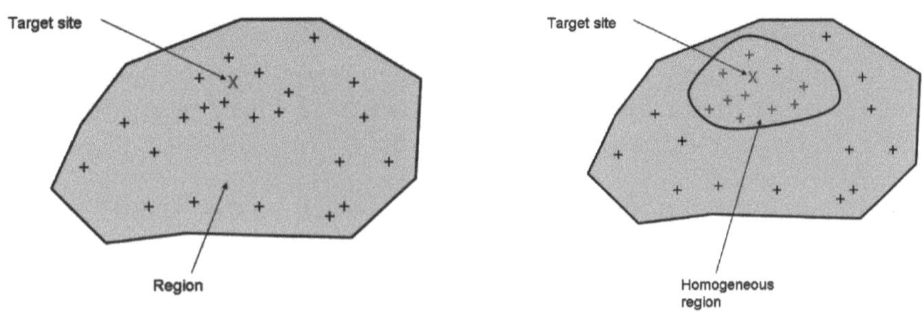

FIGURE 8.1 Illustration of regional frequency analysis regional framework.

TABLE 8.1 Selected references for the different frequency analysis frameworks.

	Univariate	Multivariate
Local	Chapter 2 and references therein	Chapters 5 and 6 and references therein
Regional	Stedinger and Tasker (1986), Burn (1990), Hosking and Wallis (1997), Wazneh et al. (2013), Haddad and Rahman (2020)	Chebana and Ouarda (2007, 2009), Chebana et al. (2009), Sadri and Burn (2012, 2014), Ben Aissia et al. (2015), Zhang et al. (2015), Requena et al. (2016), Masselot et al. (2017), Tosunoglu and Can (2016), Abdi et al. (2017), Azam et al. (2018), Šimková (2018), Ghafori et al. (2019), Vanem (2020)

are among the most commonly developed and employed. The index-flood methods assume homogeneity of the region. In other words, it assumes that all sites forming a region have the same frequency distribution apart from a scale factor related to each site. To keep the focus on the multivariate RFA framework, we present briefly the univariate RFA aspects. The reader is referred to Ouarda (2016) for an overview of RFA where a variety of methods and techniques are presented. For more clarity, Table 8.1 provides some references for the different frameworks by focusing on the multivariate regional framework.

8.2 Multivariate regional frequency analysis

Most literature related to multivariate HFA dealt with at-site (local) framework. However, little but growing number of studies deal with the multivariate HFA in *regional* framework leading to the multivariate RFA framework. On the one hand, the advantages and usefulness of considering several variables of the hydrological event, leading to the multivariate framework, are largely studied (see previous chapters). On the other hand, based on different problematic (data shortage and uncertainty) and for a different objective (estimation at ungauged sites and reduce uncertainty), the regional framework is motivated and largely studied as well. Therefore considering simultaneously both frameworks (multivariate and regional) leads to a more realistic framework combining advantages of both, even though it is more complex and requires more data and information. Indeed, the multivariate regional framework takes into account the dependence structure between variables of the hydrological event, offers more flexibility to practitioners, provides estimation at ungauged or partially gauges sites, and allows to reduce estimation uncertainty. Note that it is possible to carry out a separate *univariate* RFA for each of the variables describing the event and in particular having a variable-specific homogeneous region. Nevertheless, it is more rational to identify a single *joint* homogeneous region. This is because the analysis deals with a single event phenomenon even though described through several variables. In addition, as in the at-site case, univariate RFA can only provide limited risk evaluation at ungauged sites and it is not able to appropriately represent multiple features of the hydrological event, especially the dependence between these features.

The main steps of univariate RFA, delineation and regional estimation, remain valid in multivariate RFA. In the latter, Chebana and Ouarda (2007) proposed multivariate discordancy and homogeneity statistical tests for the delineation step. A regional estimation procedure in the multivariate RFA context is presented in Chebana and Ouarda (2009) as an extension of the index-flood model. In order to perform multivariate regional HFA, based on the multivariate index-flood model, the following procedure is recommended:

1. Detecting multivariate outliers and discordant sites by screening the data;
2. Clustering sites in homogenous regions based on the similarity of catchment descriptors;
3. Selecting the *joint* multivariate regional distribution composed of the marginal distributions as well as the copula; and
4. Using the joint distribution of the previous step, for a given return period, estimate of multivariate quantiles for a target site.

Fig. 8.2 illustrates the main steps of the multivariate RFA procedure which will be given in more detail in the next sections.

Traditional univariate RFA is well established and widely considered both in terms of applications and developments. However, the multivariate RFA has been developed recently where the discordancy and the homogeneity tests were introduced based on the multivariate L-moments. Later, the multivariate index-flood model was developed based mainly on copulas and multivariate quantile curves. Masselot et al. (2017) proposed improved versions of the multivariate (and univariate) homogeneity tests using bootstrap permutations to obtain the corresponding p-values.

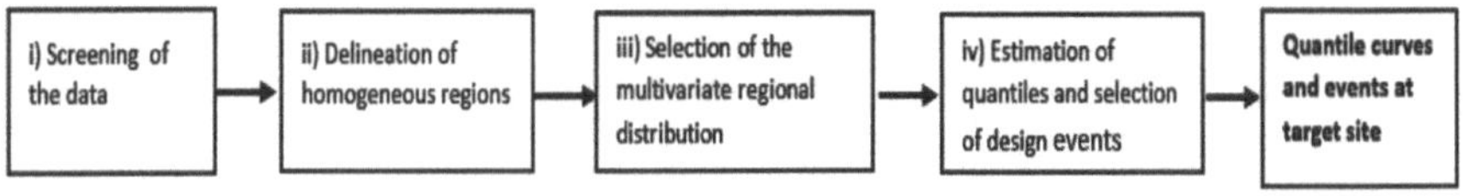

FIGURE 8.2 Main steps of multivariate regional frequency analysis.

Requena et al. (2016) presented a complete multivariate RFA study focusing on practical aspects of the whole procedure and applied it on floods. In terms of applications, a number of studies can be found, such as Chebana et al. (2009), Sadri and Burn (2012), and Azam et al. (2018) for the multivariate L-moment homogeneity test. Ben Aissia et al. (2015) considered the multivariate index-flood model on floods. Very recently, Vanem (2020) presented a bivariate RFA applied to extreme waves whereas Šimková (2018) studied a trivariate RFA on precipitations. A growing number of studies considered multivariate RFA on droughts as well (e.g., Abdi et al., 2017; Azam et al., 2018; Ghafori et al., 2019; Sadri & Burn, 2014, Tosunoglu & Can, 2016; Zhang et al., 2015).

In the following, we present the methodology related to multivariate RFA. However, in terms of applications, we suggest Requena et al. (2016) where a fully detailed case study is provided.

8.3 Delineation

The homogeneity of a region is a crucial assumption in the delineation step of a RFA. In the cases where this assumption is not fulfilled, the quality of the risk estimation at ungauged sites would be negatively impacted. Hence, it is important to consider checking regional homogeneity using statistical tests. We first present the discordancy test and then the homogeneity test. These multivariate tests are based on multivariate L-moments (briefly described in the Appendix).

8.3.1 Multivariate discordancy

Screening of the data is a preliminary step in RFA which can be performed based on two statistical tests: (a) at-site multivariate outlier detection and (b) a multivariate discordancy test. In the former, the aim is to identify for each site gross errors, incorrect measurements, or unusual observations which may have a significant impact on model fitting and, subsequently, on the quantile estimates (see Chapter 3 for more details). Discordancy test aims to detect discordant *sites* in a region since those sites affect the quality of the regional homogeneity and the final results.

Let (x_{ij}, y_{ij}) with $i = 1, \ldots, N$ and $j = 1, \ldots, n_i$ be the available bivariate data set in a region of N sites and n_i is the number of observations at site i. We consider that hydrological events are independent and are identically distributed (such as for annual floods) as commonly considered in HFA. In addition, for RFA, in order to keep the simplicity of the presentation, sites are supposed to be uncorrelated. For a given site i, let $D_i$be a matrix defined by:

$$D_i = \frac{1}{3}(U_i - \overline{U})^t S^{-1}(U_i - \overline{U}) \tag{8.1}$$

where

$$S = (N-1)^{-1}\sum_{i=1}^{N}(U_i - \overline{U})(U_i - \overline{U})^t \tag{8.2}$$

$$\overline{U} = N^{-1}\sum_{i=1}^{N} U_i \tag{8.3}$$

the matrix $U_i^t = \left[\Lambda_2^{*(i)}\Lambda_3^{*(i)}\Lambda_4^{*(i)}\right]$, A^t is the transpose of a matrix A, the three multivariate L-moment matrices $\Lambda_2^{*(i)}$, $\Lambda_3^{*(i)}$ and $\Lambda_4^{*(i)}$ defined in their general form as the matrices Λ_r^* defined with L-comoment coefficients given by

$$\tau_{2[jk]} = \frac{\lambda_{2[jk]}}{\lambda_1^{(k)}} \text{ and } \tau_{r[jk]} = \frac{\lambda_{r[jk]}}{\lambda_2^{(k)}}, \text{ for } r \geq 3 \tag{8.4}$$

$$\lambda_{r[jk]} = \mathrm{cov}\left(X^{(j)}, P_{r-1}^*\left(F_k\left(X^{(k)}\right)\right)\right) \tag{8.5}$$

where $\lambda_r^{(k)}$ is the univariate r^{th} L-moment of the variable $X^{(k)}$, and $\lambda_{r[jk]}$ are the L-comoment, $P_{r-1}^*(u)$ is the shifted Legendre polynomial, and $F(\cdot)$ is the cumulative distribution function (cdf) of the variable under consideration. The reader is also referred to Appendix for a brief presentation of L-moments.

The discordancy of a site i can be then obtained by using a norm $\|D_i\|$ of the matrixD_i. Several matrix norms can be used for this purpose. We consider the spectral norm $\|A\|$, where for a matrix A with elements a_{ij}, is defined as the square root of the maximum eigenvalue of A^tA:

$$\|A\| = \sqrt{\text{maximum eigenvalue of } A^tA} \tag{8.6}$$

In principal, if $\|D_i\|$ is large, then the corresponding site i is considered as discordant. For large regions, as a critical value for $\|D_i\|$, the constant $c = \chi_{1-0.05}(3)/3 = 2.6$ may be considered, where $\chi_{1-\alpha}(d)$ is the quantile of a chi-square distribution of order α with d degrees of freedom. However, despite not exceeding the threshold, sites having large $\|D_i\|$ values should be checked. Once detected, discordant sites could be either placed to other appropriate region or discarded from the analysis. In addition, regarding small regions, the definition of the critical value should be adapted. Based on simulations, compared to the univariate discordant tests, the $\|D_i\|$ statistics provide better indications regarding discordant sites in a heterogeneous region in particular for sites where the whole joint the distribution is concerned.

8.3.2 Multivariate homogeneity

A number of statistical tests have been developed in the literature in order to make a decision regarding the homogeneity of a set of sites composing a region. In this regard, the L-moment based test proposed by Hosking and Wallis is one of the most used and well-known, denoted as HW test in the following (Hosking and Wallis, 1997).

The multivariate statistics of the homogeneity test is given by:

$$V_{\|.\|} = \left(\left(\sum_{i=1}^{N} n_i \right)^{-1} \sum_{i=1}^{N} n_i \left\| \Lambda_2^{*(i)} - \overline{\Lambda_2^*} \right\|^2 \right)^{1/2} \tag{8.7}$$

where $n_i(i=1,\ldots,N)$ is the sample size of the ith site, and $\Lambda_2^* = \left(\sum_{i=1}^{N} n_i\right)^{-1} \sum_{i=1}^{N} n_i \Lambda_2^{*(i)}$ is the weighted mean of at-sites' $\Lambda_2^{*(i)}$ with elements given in (8.4). As for the discordancy, the spectral norm in (8.6) is one of the appropriate choices that can be applied to compute V even though any other matrix norm can be considered. The univariate V statistic of Hosking and Wallis is a particular case of $V_{\|.\|}$ when dealing with a single variable.

In order to measure the heterogeneity of a set of sites, $V_{\|.\|}$ needs to be standardized leading to the statistic given by:

$$H_{\|.\|} = \frac{V_{\|.\|} - \mu_{V\mathrm{sim}}}{\sigma_{V\mathrm{sim}}} \tag{8.8}$$

where $\mu_{V\mathrm{sim}}$ and $\sigma_{V\mathrm{sim}}$ are respectively the mean and standard deviation of a large number N_{sim} of $V_{\|.\|}$ values obtained from simulated regions. The latter should be homogeneous and similar to the observed regions in terms of number of sites as well as record length of each site.

The simulated regions are based on a bivariate distribution. The latter should be flexible and general, to include most distributions commonly used in hydrology, in order to avoid subjective selection of that distribution. Recall that a multivariate distribution is composed of margins and a copula (see Chapter 5). In the multivariate case and in the initial version of $H_{\|.\|}$, for each site, margins are simulated from a four-parameter Kappa distribution (as in the univariate setting) whereas an extreme value copula is considered to represent the dependence structure between the involved variables. Since a classical extreme value copula is not flexible enough, other copulas are considered in a more recent version of the test. To this end, a two-parameter copula is employed (Requena et al., 2016). Indeed, this category of copulas is flexible and general, where several well-known copula families are special cases, such as the copulas BB1 (Clayton and Gumbel special cases), BB6 (Joe and Gumbel), BB7 (Joe and Clayton), and BB8 (Joe and Frank) (see also Chapter 5).

Depending on the value of $H_{\|.\|}$, a decision concerning the homogeneity of the observed region can be taken. As a statistical *homogeneity test*, the homogeneity is rejected if $H_{\|.\|} > 1.64$ with a first kind error 5%. However, as a *heterogeneity measure*, if either $H_{\|.\|} < 1$, $1 < H_{\|.\|} < 2$, or $H_{\|.\|} > 2$, then the corresponding region can be considered respectively as homogeneous, possibly homogenous, or definitely heterogonous.

Even though the univariate Hosking–Wallis test is widely used and has a good power, it is affected by some factors. First, it requires to fit or to assume a parametric distribution to the data (hence it is considered as a parametric L-moment homogeneity test). This extra step, not common in statistical testing, may lead to distribution misspecification as well as to increase the uncertainty generated from the estimation of its parameters. Second, the threshold on which a decision to accept or to reject the homogeneity is not well justified. In the multivariate setting, the effects of these limitations, especially the first one, could become more significant (more parameters to be estimated and more distributions to be selected). Recently, Masselot et al. (2017) proposed a nonparametric version of the parametric L-moment homogeneity test. In the former, simulated homogeneous regions are generated by randomly permuting the pooled data between sites, hence avoiding to fit or assume a parametric distribution. Regarding the decision rule, the proposed one is based on the p-value instead of a rejection threshold. The p-value is obtained using resampling methods (permutation, bootstrap, and Pólya). Besides their simplicity, the permutation and bootstrap methods are more powerful than the parametric test (the permutation method leads to the most powerful test). Explicitly, for each simulated region, the corresponding statistic $V^{(j)}_{\|.\|}$, $j = 1, \ldots, N\text{sim}$, is computed over the Nsim number of simulations. Hence, the p-value can obtained as $p - \text{value}_H = \#\left\{V^{(j)}_{\|.\|} > V_{\|.\|}\right\}/N\text{sim}$. Therefore, given a significance level α, the homogeneity as a null hypothesis is rejected if $p\text{-value}_H < \alpha$. The nonparametric test avoids to establish a threshold of the statistic to make a decision.

Fig. 8.3 provides an overview of the delineation step.

In the RFA framework, a variety of methods are available in order to form groups of sites that are potentially homogeneous regions, such as canonical correlation analysis, region of influence, and C-means (e.g., Ouarda, 2016 for a relatively recent overview). These methods consider

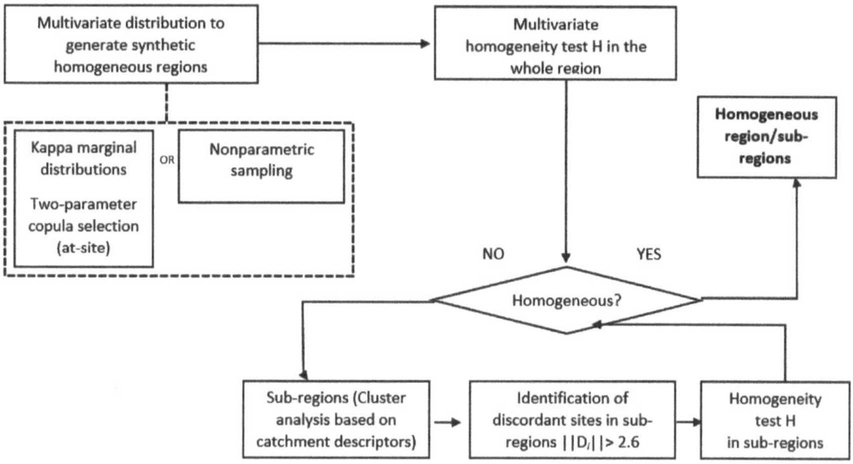

FIGURE 8.3 Summary of the delineation step.

several variables describing different features of the sites when obtaining appropriate clusters, including geographic (e.g., latitude and longitude), physiographic (e.g., drainage basin area, elevation, basin slope, river slope), climatic (e.g., annual mean of total precipitation), or hydrological descriptors (e.g., flood quantiles of given return periods). For a reliable analysis, the size of the subregions (clusters) should be balanced where a large number of sites are needed for an appropriate quantile estimates, whereas a small number of sites are more likely to ensure homogeneity.

If the homogeneity of a region is rejected, even after checking the discordant sites, then in principle no unique bivariate distribution can be selected to model the flood (or other events) response in that region. Consequently, one of the options is to divide the region with appropriate clustering method into smaller subregions provided that at least they contain enough number of sites. In the newly obtained subregions, the discordancy of the sites is tested again and the statistic H is evaluated as well. In the case one or more subregions remain heterogenous, then the clustering procedure can be repeated by considering different variables or number of clusters. Note that the homogeneity of a region is not always guaranteed. Fig. 8.3 provides an overview of the delineation step within RFA.

In the multivariate setting, where several variables describe the hydrological phenomenon, one can argue that it is easier and preferred to test the multivariate homogeneity based on a series of univariate tests (one for each involved variable) instead of a single multivariate test. However, in general statistical testing, using multiple tests increases the significance level (which should be low, usually 5%). Indeed, considering d independent univariate tests, each of which is at the 5% significance level, then $(1 - 0.95^d)$ is the probability of getting at least one significant result, which may be unacceptably large. This is not the case by using a single multivariate test. In addition, with a multivariate test, the correlation between variables is taken into account.

8.4 Multivariate estimation (index-flood model)

The initial and univariate index-flood model was introduced by Dalrymple (1960) to deal with floods in RFA. However, other hydrological events, such as storms and droughts, can also be treated with similar forms of this model. One of the main assumptions of this model is the homogeneity of the region to be studied. In other words, apart from a scale parameter that distinguishes each site, all sites in the region should have the same frequency distribution. In the univariate case, given a region with N sites, the index-flood model is expressed as follows:

$$Q_i(p) = \mu_i q(p), \quad i = 1, \ldots, N \text{ and } 0 < p < 1 \tag{8.9}$$

where $Q_i(p)$ is the quantile associated to the nonexceedence probability p at site i, μ_i is the index-flood, and $q(.)$ is the regional growth curve.

The index-flood parameter μ_i is considered as a location parameter of the distribution of site i, and hence it can be estimated, for instance, as the sample mean or median at each site i. To estimate the growth curve $q(.)$, the data of each site in the region should be standardized and then pooled to fit an appropriate distribution (as in the usual univariate HFA). More information regarding the univariate index-flood model can be found in Hosking and Wallis (1997), and a recent review in Wazneh et al. (2013).

The multivariate index-flood model, as an extension of (8.9), is based on two main concepts: the notions of copulas and multivariate quantile curves. As described in Chapter 5, copulas are mainly utilized in order to model the dependence structure between the variables involved to describe the hydrological phenomenon, such as peak and volume for floods. Recall that the multivariate quantile version considered in bivariate HFA is a curve composed of combinations of the involved variables corresponding to the same risk level (see Chapter 6 or Chebana and Ouarda, 2011). The bivariate index-flood model is formulated for a given nonexceedance probability $p \in (0, 1)$ at a gauged site i as follows:

$$Q^i_{X,Y}(p) = \begin{bmatrix} \mu^i_X \\ \mu^i_Y \end{bmatrix} q_{X,Y}(p), \quad i = 1, \ldots, N \tag{8.10}$$

where μ_X^i is the index flood of X (similarly for μ_Y^i), $Q_{X,Y}^i(p)$ is the bivariate quantile curve of X and Y for a given (simultaneous nonexceedance) probability p at site i, and $q_{X,Y}(p)$ is the dimensionless bivariate regional quantile curve. As in the univariate case, it can be seen as a "regional quantile curve" and can be obtained from the standardized and pooled data of the region. One of the main differences between the univariate and bivariate index-flood models is the nature of the corresponding quantiles, where in the former it is a single value, whereas in the latter it is rather a curve.

In order to estimate the quantile of interest at the (ungauged) target-site ℓ, given a set of N gauged sites with record length n_i at site i, $i = 1,\ldots, N$, the bivariate RFA procedure is summarized as follows:

1. Identify the region (as a set of sites) to be used in the next steps:
   a. Identify discordant sites to be potentially removed from the region by applying the multivariate discordancy test $\|D_i\|$,
   b. Apply the multivariate homogeneity test $H_{\|.\|}$ to the region of the remaining sites (from step 1a). As a notation, these sites are indexed from 1 to N' (with $N' \leq N$).
2. For each site i, $i = 1,\ldots, N'$ in the homogenous region (obtained from step 1):
   a. Assess the location parameters $\hat{\mu}_{i,X}$ and $\hat{\mu}_{i,Y}$ (assumed non-null),
   b. Standardize the pooled sample of that region (x_{ij}, y_{ij}) using the location parameters in step 2a to get $\left(x'_{ij} = x_{ij}/\hat{\mu}_{i,X}, y'_{ij} = y_{ij}/\hat{\mu}_{i,Y}\right)$,
3. Select an appropriate multivariate distribution to fit the standardized data (x'_{ij}, y'_{ij}) for $j = 1, \ldots, n_i$ and $i = 1, \ldots, N'$. Note that a multivariate distribution is composed of univariate marginal distributions for each variable as well as a copula for their dependence structure (see goodness-of-fit tests for copula and margins in Chapter 5),
4. Let $\theta_1, \ldots, \theta_s$ denote the parameters of the selected multivariate distribution in step 3. Estimate the parameters of the distribution obtained in step 3:
   a. Using the standardized data of the ith gauged site, obtain an estimator $\hat{\theta}_k^{(i)}$ of the kth parameter, $k = 1,\ldots, s$ and $i = 1,\ldots, N'$ (see estimation methods in Chapter 5).
   b. Combine the estimators in step 4.a using the record length of each site n_i as

$$\hat{\theta}_k^{(R)} = \frac{\sum_{i=1}^{N'} n_i \hat{\theta}_k^{(i)}}{\sum_{i=1}^{N'} n_i}, \quad k = 1, \ldots, s \tag{8.11}$$

to obtain the weighted regional parameter estimators.

5. For a given value of p, based on the multivariate distribution from steps 3 and 4, obtain the associated bivariate quantile curve of the growth curve $\hat{q}_{x,y}(p)$. Combinations or sections of the cuvre can also be obtained from this quantile curve (see Chapter 6).
6. For a target-site ℓ, estimate the corresponding indices $\hat{\mu}_{\ell,X}$ and $\hat{\mu}_{\ell,Y}$. Unlike those in step 2, for an ungauged site, $\hat{\mu}_{\ell,X}$ and $\hat{\mu}_{\ell,Y}$ can be obtained from meteorological and physiographical features of the ungauged site. Usually, this can be obtained based on a linear regression model (see below).
7. In order to obtain an estimate of the quantile curve corresponding to the target-site ℓ, multiply component-wise each growth curve combination (step 5) by the location parameter of the target-site ℓ, $\hat{\mu}_{\ell,X}$ and $\hat{\mu}_{\ell,Y}$ (step 6):

$$\left(\hat{Q}_{x,y}(p)\right)_{\ell} = \begin{pmatrix} \hat{\mu}_{\ell,X} \\ \hat{\mu}_{\ell,Y} \end{pmatrix} \hat{q}_{x,y}(p) \quad 0<p<1 \tag{8.12}$$

An overview of the whole multivariate RFA procedure based on the multivariate index-flood model is given in Fig. 8.4.

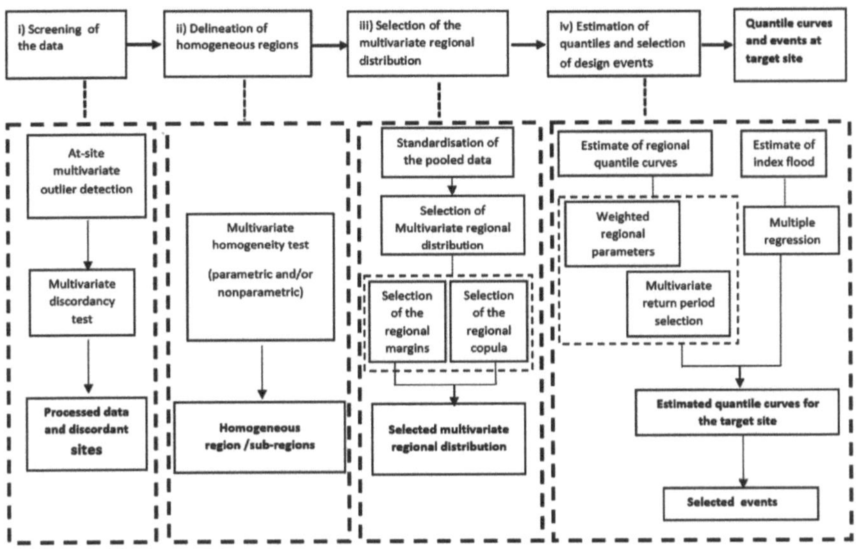

FIGURE 8.4 Summary of the complete procedure of a multivariate (bivariate) regional procedure based on the index-flood model. Source: *Adapted from Requena, A. I., Chebana, F., & Mediero, L. (2016). A complete procedure for multivariate index-flood model application.* Journal of Hydrology, 535, *559–580.*

8.5 Discussion

In the index-flood model, the index μ plays an important role, such as for standardizing the data set at each gauged site. For gauged sites, it is obtained directly from the hydrological data, as a location parameter (in the present study the sample mean is considered). However, for ungauged sites, in the absence of hydrological data, μ should be estimated based on a model as well as meteorological and physiographical features of the ungauged sites. Multiple regression has been shown to be an appropriate model where the best features can be selected. More details such as variable selection (catchment descriptors), linearity, and model fitting in this context are presented in Requena et al. (2016).

Even though the bivariate and univariate index-flood model versions are similar in terms of the main steps and principles, there are some differences. The first difference lies in the notion of quantiles as discussed earlier (real-valued vs curve, Chapter 6). In addition, the univariate index-flood model implicitly considers the nonexceedance event $(X \leq x)$. By analogy, the simultaneous nonexceedance event $\{X \leq x, Y \leq y\}$ is considered in the presented bivariate version. However, as presented in Chapter 6, other events of interest, for example, simultaneous exceedance $\{X > x, Y > y\}$, could be considered. This extension requires appropriate adaptation to the format of the presented model.

In general, both bivariate and univariate index-flood models have overall similar performances. However, in the case where the region is close to homogeneity, the bivariate model outperforms the univariate one. Indeed, small variations of the record length of sites or the region size have negligible effects on the performance of the bivariate model. However, these factors impact the whole regionalization procedure.

On the one hand, for simplicity, in the previous sections, the bivariate RFA framework is presented. On the other hand, extensions to higher d-dimensional of all the presented tests and models are available. However, some challenges may arise in practice when passing to higher dimensions. Some of them are introduced in previous chapters (e.g., Chapter 5). For instance, the number of parameters to be estimated grows quickly with the dimension d and numerical/computational difficulties become even more important for $d > 2$. Another key issue is related to copula modeling for higher dimensions. Indeed, well-known classes of copulas, such as Archimedean or extreme value copulas, are available for $d > 2$. However, they have limited flexibility for complex dependence structures (such as asymmetric dependence or different nature of dependencies). Considering other kinds of copulas, for example, Vine Copulas, is still an evolving research topic in statistics and in hydrology (see Chapter 5).

Studying the uncertainty in the multivariate local (at-site) HFA is important and a number of recent studies dealt with it. One of the related issues is the need of a long data series in order to obtain reliable quantile estimates and quantile curves. In addition to the uncertainty of the local estimate generated from the estimates of the copula and marginal distributions, in multivariate RFA, another source of uncertainty is the one related to the estimate of the index flood μ.

References

Abdi, A., Hassanzadeh, Y., Talatahari, S., Fakheri-Fard, A., & Mirabbasi, R. (2017). Regional bivariate modeling of droughts using L-comoments and copulas. *Stochastic Environmental Research and Risk Assessment, 31*(5), 1199–1210.

Azam, M., Park, H. K., Maeng, S. J., & Kim, H. S. (2018). Regionalization of drought across South Korea using multivariate methods. *Water, 10*(1), 24.

Ben Aissia, A., Chebana, F., Ouarda, T. B. M. J., Bruneau, P., Barbet, M., & Équipement, H. Q. (2015). Bivariate index flood model for a northern case study. *Hydrological Sciences Journal, 60*, 247–268.

Burn, D. H. (1990). Evaluation of regional flood frequency analysis with a region of influence approach. *Water Resources Research, 26*, 2257–2265.

Chebana, F., & Ouarda, T. B. M. J. (2007). Multivariate *L*-moment homogeneity test. *Water Resources Research, 43*, W08406. Available from https://doi.org/10.1029/2006WR005639.

Chebana, F., & Ouarda, T. B. M. J. (2009). Index flood-based multivariate regional frequency analysis. *Water Resources Research, 45*, W10435.

Chebana, F., & Ouarda, T. B. M. J. (2011). Multivariate quantiles in hydrological frequency analysis. *Environmetrics (London, Ont.), 22*, 63–78.

Chebana, F., Ouarda, T. B. M. J., Bruneau, P., Barbet, M., El Adlouni, S., & Latraverse, M. (2009). Multivariate homogeneity testing in a northern case study in the province of Quebec, Canada. *Hydrological Processes, 23*, 1690–1700.

Dalrymple, T. (1960). Flood frequency methods. *United States Geological Survey Water-Supply Paper* (1543 A), 11–51.

Ghafori, V., Sedghi, H., Sharifan, R. A., & Nazemosadat, S. M. J. (2019). Regional frequency analysis of droughts using Copula functions (case study: Part of semiarid climate of Fars province, Iran). *Iranian Journal of Science and Technology, Transactions of Civil Engineering*, 1–13.

Haddad, K., & Rahman, A. (2020). Regional flood frequency analysis: Evaluation of regions in cluster space using support vector regression. *Natural Hazards, 102*(1), 489–517.

Hosking, J. R. M., & Wallis, J. R. (1997). *Regional frequency* analysis: An approach based on L-moments. Cambridge University Press, 240 pages.

Masselot, P., Chebana, F., & Ouarda, T. B. (2017). Fast and direct nonparametric procedures in the L-moment homogeneity test. *Stochastic Environmental Research and Risk Assessment, 31*(2), 509–522.

Ouarda, T. B. M. J. (2016). *Regional flood frequency modeling. Chap, 77*, 77, 1.

Requena, A. I., Chebana, F., & Mediero, L. (2016). A complete procedure for multivariate index-flood model application. *Journal of Hydrology, 535*, 559–580.

Sadri, S., & Burn, D. H. (2012). A fuzzy C-means approach for regionalization using a bivariate homogeneity and discordancy approach. *Journal of Hydrology, 401*(3–4), 231–239.

Sadri, S., & Burn, D. H. (2014). Copula-based pooled frequency analysis of droughts in the Canadian Prairies. *Journal of Hydrologic Engineering, 19*(2), 277–289.

Šimková, T. (2018). L-moment homogeneity test in trivariate regional frequency analysis of extreme precipitation events. *Meteorological Applications, 25*(1), 11–22.

Stedinger, J. R., & Tasker, G. (1986). Regional hydrologic analysis, 2, Model-error estimators, estimation of sigma and log Pearson type 3 distributions. *Water Resources Research, 22*, 1487–1499.

Tosunoglu, F., & Can, I. (2016). Application of copulas for regional bivariate frequency analysis of meteorological droughts in Turkey. *Natural Hazards, 82*(3), 1457–1477.

Vanem, E. (2020). Bivariate regional extreme value analysis for significant wave height and wave period. *Applied Ocean Research, 101*102266.

Wazneh, H., Chebana, F., & Ouarda, T. B. (2013). Depth-based regional index-flood model. *Water Resources Research, 49*(12), 7957–7972.

Zhang, Q., Qi, T., Singh, V. P., Chen, Y. D., & Xiao, M. (2015). Regional frequency analysis of droughts in China: A multivariate perspective. *Water Resources Management, 29*(6), 1767–1787.

APPENDIX

A

Ties in the copula-based framework and in hydrology

In the majority of the statistical procedures considered in hydrological frequency analysis (HFA), it is assumed that the variables for study are continuous. For instance, dependence measures such as Kendall's tau and Spearman's rho as defined for continuous data in general do not apply directly to discrete data. Nevertheless, in practice, for a number of reasons, such as rounding and lack of measurement precision, it is possible to observe identical values of variables (called ties) in a dataset arising from the observation of a naturally continuous random variable. The presence of ties is not desirable for copula-based statistical analysis, in which ranks are important. Let $X_1, \ldots, X_n$ be a univariate sample from a distribution function F. If $X_i = X_j$ for some *i different from j*, then the sample contains *ties*. In the case where F is continuous, ties occur with probability zero. In the multivariate framework, a d-dimensional dataset $\mathbf{X}_1, \ldots, \mathbf{X}_n$ contains *ties* if at least one component univariate sample $X_{1j}, \ldots, X_{nj}, j \in \{1, \ldots, d\}$ contains ties (e.g., Hofert et al., 2018).

As in other fields, ties can be present in a number of hydrological phenomena and events. However, a limited number of studies are available in hydrological applications dealing with ties, and especially in the multivariate setting (e.g., Salvadori & De Michele, 2006). As an example, as a result of measuring the volume of precipitation in multiples of 0.1 mm, continuous variables will experience some degree of discretization and ties are introduced (Vandenberghe et al., 2010). Droughts are another interesting hydrological example where ties could be present, in particular in a multivariate framework since ties may be present in drought duration and drought interarrival time variables.

Many statistical tools can be strongly affected if the presence of ties is ignored, whereas a few of them still can either provide meaningful results or be adapted for ties. For instance, the presence of ties may have serious effects on the estimation of Kendall's tau and Spearman's tau coefficients. In addition, copula parameter estimation is affected especially in terms of bias when ties are present but not in terms of the variability. Furthermore, the presence of ties has an important effect on the goodness-of-fit tests for copula (Hofert et al., 2018).

To overcome the negative effects of ties on the statistical procedures and the obtained results, several modified versions of the dependence measures have been developed in the literature (Denuit & Lambert, 2005). To carry out inference using goodness-of-fit tests in the presence of ties, there are mainly two possible approaches: (1) transform the original data set so that the resulting multivariate scaled ranks are free of ties or (2) adapt the inference procedures so that they give meaningful results in the presence of ties. More details can be found in Hofert et al. (2018).

References

Denuit, M., & Lambert, P. (2005). Constraints on concordance measures in bivariate discrete data. *Journal of Multivariate Analysis, 93*, 40–57.

Hofert, M., Kojadinovic, I., Mächler, M., & Yan, J. (2018). *Elements of copula modeling with R*. Springer International Publishing.

Salvadori, G., & De Michele, C. (2006). Statistical characterization of temporal structure of storms. *Advanced Water Resources, 29*(6), 827–842.

Vandenberghe, S., Verhoest, N. E. C., & De Baets, B. (2010). Fitting bivariate copulas to the dependence structure between storm characteristics: a detailed analysis based on 105 year 10 min rainfall. *Water Resources Research, 46*, W01512. Available from https://doi.org/10.1029/2009WR007857.

APPENDIX

B

Statistical depth functions

In the present appendix, some necessary background elements related to depth functions are provided. Depth functions are the basis of some statistical tools and techniques discussed in Chapter 3, regarding descriptive analysis, and for some tests presented in Chapter 4.

Given the lack of a natural ordering or ranking notion in the multivariate framework, the main aim of introducing depth functions was to define multivariate extensions of the rank and order notions. To this end, Tukey (1975) first presented depth functions by proposing the half-space depth function. Later, several types of depth functions were defined and then standardized and classified by Zuo and Serfling (2000). Depth functions have been considered in a variety of applications such as in econometric and social studies, industrial quality control, and in particular in hydrology such as by Chebana and Ouarda (2008), Bárdossy and Singh (2008), Chebana and Ouarda (2011), and Singh et al. (2016). Specifically, depth functions are applied in hydrological frquency analysis by Chebana and Ouarda (2011) as well as in regional flood frequency analysis by Wazneh et al. (2016) among other studies.

A depth function $D(x; F)$, defined for a given multivariate cumulative distribution function F on $R^d(d \geq 1)$ and x in R^d, is any bounded and nonnegative function that meets the following properties:

1. *Affine invariance*: The depth of a point $x \in R^d$ should be the same even after any linear transformation, that is, $D(x; F_X) = D(Ax + b; F_{AX+b})$ for any random vector X in R^d, any $d \times d$ nonsingular matrix A, and any d-vector b.
2. *Maximality at center*: For a distribution having a uniquely defined center, the depth function should attain its maximum value at this center (*deepest point*).

3. *Monotonicity relative to deepest point*: As a point $x \in R^d$ moves away from the deepest point along any fixed ray through the center, the depth at x should decrease monotonically.
4. *Vanishing at infinity*: The depth of a point x should be close to zero as the corresponding norm $\|x\|$ approaches infinity.

A number of depth functions are proposed in the literature of which the following are some of the most used ones:

Tukey depth (called also the half-space depth): It is given by:

$$HD(x; P) = \inf\{P(H) : H \text{ is a closed half-space that contains } x\} \quad \text{for } x \in R^d \tag{B.1}$$

for a probability P on R^d.

The empirical *half-space* depth function is defined by replacing the probability function $P(H)$ with the proportion of sample observations falling into a half-space H.

Oja depth (called also the Simplicial volume depth): It is expressed as:

$$SVD(x, F) = (1 + E[\Delta(S_n[x, X_1, \ldots, X_d])])^{-1} \quad \text{for } x \in R^d \tag{B.2}$$

where $\Delta(S_n[x, x_1, \ldots, x_d])$ is the volume of the closed *d-simplex* $S_n[x, x_1, \ldots, x_d]$ formed by the points $x, x_1 \ldots, x_d \in R^d$. A d-simplex is defined as the convex hull of these points which can be seen as a generalization of triangles in d-dimension.

Mahalanobis depth: It is the simplest one and is based on the *Mahalanobis* distance:

$$d_A^2(x, y) = (x - y)' A^{-1} (x - y) \quad \text{for } x, y \in R^d \tag{B.3}$$

where A is any semi-definite-positive matrix. The Mahalanobis depth, noted MD, is formulated as:

$$MD(x, F) = \left(1 + d_{A(F)}^2(x, \mu(F))\right)^{-1} \quad \text{for } x \in R^d \tag{B.4}$$

where F is a multivariate distribution with a scale measure *A(F)* and a location parameter $\mu(F)$,

Liu depth (called also the Simplicial depth): The *Simplicial depth,* noted SD, of a point x in R^d with respect to a distribution F is given by

$$SD(x, F) = P_F\{x \in S_n[X_1, \ldots, X_{d+1}]\} \quad \text{for } x \in R^d \tag{B.5}$$

where $X_i \sim F, i = 1, \ldots, d + 1$, and S_n is as defined above closed *d-simplex* $S_n[x, x_1, \ldots, x_d]$ formed by the points $x, x_1 \ldots, x_d \in R^d$.

In terms of computations, the Mahalanobis depth is among the simplest ones to evaluate. However, for other depth functions, it requires specific algorithms. Algorithms for the computation of the half-space and the Simplicial depth functions are available in the literature.

References

Bárdossy, A., & Singh, S. K. (2008). Robust estimation of hydrological model parameters. *Hydrology and Earth System Sciences, 12*(6), 1273–1283.

Chebana, F., & Ouarda, T. B. (2008). Depth and homogeneity in regional flood frequency analysis. *Water Resources Research, 44*(11).

Chebana, F., & Ouarda, T. B. (2011). Depth-based multivariate descriptive statistics with hydrological applications. *Journal of Geophysical Research: Atmospheres, 116*(D10).

Singh, S. K., McMillan, H., Bárdossy, A., & Chebana, F. (2016). Nonparametric catchment clustering using the data depth function. *Hydrological Sciences Journal, 61*(15), 2649–2667.

Tukey, J. (1975). Mathematics and picturing data. In *Proceedings of the International 17 Congress of Mathematics*, 2, 523–531.

Wazneh, H., Chebana, F., & Ouarda, T. B. (2016). Identification of hydrological neighborhoods for regional flood frequency analysis using statistical depth function. *Advances in Water Resources, 94*, 251–263.

Zuo, Y., & Serfling, R. (2000). General notions of statistical depth function. *The Annals of Statistics, 28*(2), 461–482.

APPENDIX

C

Multivariate *L*-moments

The *L*-moments, introduced as an alternative to traditional moments, offer strong advantages for the modeling of heavy-tailed distributions such as some of the distributions used in hydrological frequency analysis. The properties and advantages of *L*-moments are presented in detail, for instance, by Hosking and Wallis (1997).

L-moments are linear combinations of order statistics sampled from an underlying distribution. The univariate *L*-moment of order r is defined as:

$$\lambda_r = \int_0^1 F^{-1}(u) P^*_{r-1}(u) du \tag{C.1}$$

where $P^*_{r-1}(u)$ is the shifted Legendre polynomial and $F(\cdot)$ is the cumulative distribution function of the variable of interest.

L-moments have been extended to the multivariate case by Serfling and Xiao (2007). They have been considered in hydrometeorology to define multivariate tests (Chebana & Ouarda, 2007), to estimate copula parameters (Brahimi et al., 2015), and to establish goodness-of-fit tests for multiparameter copulas (Ben Nasr & Chebana, 2019). In the following, the bivariate *L*-moment case is briefly presented.

Let $X^{(1)}$ and $X^{(2)}$ be random variables with distributions F_1 and F_2, respectively. By analogy with a covariance representation of *L*-moments of order $k \geq 1$, multivariate *L*-moments are matrices Λ_k with *L*-comoment elements defined for $\left(X^{(1)}, X^{(2)}\right)$ by:

$$\lambda_{k[12]} = \mathrm{Cov}\left(X^{(1)}, P^*_{k-1}\left(F_2(X^{(2)})\right)\right), \quad k = 2, 3, \ldots \tag{C.2}$$

In an analogous way, we can define $\lambda_{k[21]}$. Note that, unlike a covariance matrix, the elements $\lambda_{k[ij]}$ and $\lambda_{k[ji]}$ are not necessarily equal. Also note that for positive b and d, and arbitrary a and c, we have:

$$\lambda_{k[12]}(a + bX^{(1)}, c + dX^{(2)}) = b\lambda_{k[12]}(X^{(1)}, X^{(2)}) \tag{C.3}$$

Particularly, the first L-comoment elements are ($k = 2, 3$, and 4):

$$\begin{aligned} \lambda_{2[12]} &= 2\mathrm{Cov}\left(X^{(1)}, F_2(X^{(2)})\right) \\ \lambda_{3[12]} &= 6\mathrm{Cov}\left(X^{(1)}, \left(F_2(X^{(2)}) - 1/2\right)^2\right) \\ \lambda_{4[12]} &= \mathrm{Cov}\left(X^{(1)}, 20\left(F_2(X^{(2)}) - 1/2\right)^3 - 3\left(F_2(X^{(2)}) - 1/2\right) + 1\right) \end{aligned} \tag{C.4}$$

which, respectively, represent L-covariance, L-coskewness, and L-cokurtosis.

Multivariate L-moments capture the behavior and the attractive properties of univariate L-moments (Serfling & Xiao, 2007). As in the univariate case, to summarize a distribution, it is convenient to define dimensionless versions of L-moments. To this end, we introduce the L-moment ratios or the L-comoment coefficients given by:

$$\tau_{k[12]} = \frac{\lambda_{k[12]}}{\lambda_2^{(1)}}, \quad \text{for } k \geq 3 \text{ and } \tau_{2[12]} = \frac{\lambda_{2[12]}}{\lambda_1^{(1)}} \tag{C.5}$$

where $\lambda_k^{(1)} = \lambda_{k[11]}$ is the classical kth L-moment of the variable $X^{(1)}$.

The matrix of the L-comoment coefficients is written as:

$$\Lambda_k^* = \left(\tau_{k[ij]}\right)_{i,j=1,2} = \begin{pmatrix} \tau_{k[11]} & \tau_{k[12]} \\ \tau_{k[21]} & \tau_{k[22]} \end{pmatrix}. \tag{C.6}$$

Particularly, for $k = 2$, the L-covariation matrix is given by:

$$\Lambda_2^* = \begin{pmatrix} \tau_{2[11]} & \tau_{2[12]} \\ \tau_{2[21]} & \tau_{2[22]} \end{pmatrix} \tag{C.7}$$

and for $k = 1$, the first-order bivariate L-moment corresponds to the mean vector $\lambda_1 = E\left(X^{(1)}, X^{(2)}\right)^t$.

The multivariate L-moments defined previously are based on a theoretical population distribution. However, their finite-sample versions are useful to define statistical tests and also to estimate multivariate distribution parameters. Their formulas and properties are presented by Serfling and Xiao (2007).

References

Ben Nasr, I., & Chebana, F. (2019). Multivariate L-moment based tests for copula selection, with hydrometeorological applications. *Journal of Hydrology, 579*, 124151.

Brahimi, B., Chebana, F., & Necir, A. (2015). Copula representation of bivariate L-moments: a new estimation method for multiparameter two-dimensional copula models. *Statistics, 49*(3), 497–521.

Chebana, F., & Ouarda, T. B. (2007). Multivariate L-moment homogeneity test. *Water Resources Research, 43*(8).

Hosking, J. R. M., & Wallis, J. R. (1997). *Regional frequency analysis* (p. 240) Cambridge University Press.

Serfling, R., & Xiao, P. (2007). A contribution to multivariate L-moments: L-comoment matrices. *Journal of Multivariate Analysis, 98*(9), 1765–1781.

APPENDIX

D

p-Value computation

The notion of p-value is important in hypothesis testing. It is defined as the probability, calculated under the null hypothesis, of having an outcome as extreme as the observed value in the sample. It is commonly used in practice as a simple criterion to decide for the acceptance or rejection of the null hypothesis. The evaluation of the p-value is based on the distribution of the test statistic. According to the complexity of the test statistic, this distribution can be known, asymptotic or approximated. For some of the considered tests (such as those in Chapters 4 and 5), the asymptotic or the exact distribution of test statistic is unknown or difficult to obtain. Consequently, approximations of the distribution of test statistic, generically denoted S, under the null hypothesis are considered. To this end, resampling methods are used, including permutation of Fisher and bootstrap methods. They are briefly described below, whereas more details can be found in Good (2005).

The permutation method consists in permuting a large number of permutations n_p times the sample $(x_i)_{i=1,\ldots,n}$ *without replacement*. For each permuted sample, the first s elements constitute the first subsample and the remaining constitutes the second one. The test statistic S is calculated for each permutation $(S^*_{i,i=1,\ldots,n_p})$. Suppose the null hypothesis is to be rejected if the test statistic S is smaller than some threshold (this is usually the case for a large number of situations). Then, the p-value is the proportion among $(S^*_{i,i=1,\ldots,n_p})$ that are smaller or equal to the value S_{obs} corresponding to S obtained from the original observed sample. To apply the permutation test of Fisher method, the observations should be exchangeable, that is, the observations are independent and identically distributed. The bootstrap method is similar to the permutation test of Fisher, except that the sample $(x_i)_{i=1,\ldots n}$ is resampled *with replacement* and the independence assumption is necessary. In the context of hydrological frequency analysis, Masselot et al. (2017) applied these techniques to develop nonparametric homogenity and discordancy multivariate (and univariate) tests for regional frequency analysis.

Reference

Good, P. (2005). *Multivariate analysis. Permutation, parametric and bootstrap tests of hypotheses* (pp. 169–188). Springer.

Masselot, P., Chebana, F., & Ouarda, T. B. (2017). Fast and direct nonparametric procedures in the L-moment homogeneity test. *Stochastic Environmental Research and Risk Assessment, 31*(2), 509–522. https://doi.org/10.1007/s00477-016-1248-0.

Index

Note: Page numbers followed by "*f*" and "*t*" refer to figures and tables, respectively.

D

E

K

L

M